TIGER BEETLES

of the **SOUTHEASTERN UNITED STATES**

TIGER BEETLES

of the SOUTHEASTERN UNITED STATES

A Field Guide

Giff Beaton, R. Stephen Krotzer & Brian D. Holt

The University of Alabama Press Tuscaloosa

The University of Alabama Press
Tuscaloosa, Alabama 35487-0380
uapress.ua.edu

Typeface: Scala Pro

Manufactured in Korea

Cover images: Florida Metallic Tiger Beetle (*Tetracha floridana*), photograph by Giff Beaton; size comparison of southeastern US tiger beetles, photograph by R. Stephen Krotzer

Cover design: Michele Myatt Quinn

Cataloging-in-Publication data is available from the Library of Congress.
ISBN: 978-0-8173-5998-0
E-ISBN: 978-0-8173-9339-7

This book is dedicated to Paul "Skip" Choate (1948–2012), entomologist, beetle lover, discoverer of the Highlands Tiger Beetle (*Cicindelidia highlandensis*), mentor, and wonderful all-around human being. He was always asking questions and pushing others to do the same, and he was literally the inspiration for this guide. He was an amazing man, and his humor and dedication to his passions, including tiger beetles, are sorely missed.

Contents

Acknowledgments

This project would not have been possible without the assistance of many institutions and individuals. We would like to thank all those who helped track down obscure records or references, opened their collections to the authors for study, shared their personal records from the southeastern states, or otherwise made this book much more useful and complete:

Alabama Alabama Department of Conservation and Natural Resources, Natural Heritage Section (Wayne Barger, Ashley S. Peters); Anniston Museum of Natural History (Dan Spaulding); Auburn University (Charles Ray); Jacksonville State University (Lori Tolley-Jordan); Troy University (Alvin Diamond); University of Alabama (G. Milton Ward); University of North Alabama (Paul Kittle); John Bloch; and Mary Jane Krotzer.

Florida Florida State Collection of Arthropods (Paul Skelley, director); Paynes Creek State Historical Park (Jackson Mosley); Ken Allen; Lyn and Brooks Atherton; Dave Brzoska; Skip and Angie Choate; Pierson Hill; Barry Knisley; Ted MacRae; Jonathan Mays; Roy Morris; and Jeff Slotten.

Georgia Columbus State University (George Stanton, Harlan Hendricks); Georgia College (Chris Skelton); Georgia Southern University (Lance Durden); Georgia Southwestern State University (Ian Brown, Hap Tietjen); North Georgia College (Karrie Anne Fadroski); South Georgia State College (A. G. Cook); University of Georgia (Rick Hoebeke, Cecil Smith); Valdosta State University (Mark Blackmore); Whit Brown; Marion Dobbs; Bruce Hallett; John Jensen; Roy Morris; and Dirk Stevenson.

Mississippi Mississippi Department of Wildlife, Fisheries, and Parks Museum of Natural Science (Scott Peyton); Mississippi Entomological

Museum (Richard Brown, Terry Schiefer, JoVonn Hill); University of Mississippi (Paul Lago); University of Southern Mississippi (Donald Yee); and Robert Ward.

North Carolina National Park Service/Great Smoky Mountains National Park (Becky Nichols, Paul Super); North Carolina Natural Heritage Program (Steve Hall, Harry LeGrand, Ed Corey, Tom Howard); North Carolina State University (Bob Blinn, Dave Stephan); Chris Hill; Dave Melius; and Jeff Pippen.

South Carolina Joey Holmes.

Tennessee Tennessee Department of Environment and Conservation; National Park Service/Great Smoky Mountains National Park; University of Tennessee (David Etnier, Jennifer Parris Joice); University of Tennessee at Chattanooga (Stylianos Chatzimanolis); Dan Duran; and Robert Ward.

Other institutions and colleagues who provided valuable support to the authors include Carnegie Museum of Natural History (Bob Davidson, Bob Acciavatti); Colorado State University C. P. Gillette Museum of Arthropod Diversity (Boris C. Kondratieff); Smithsonian Institution (Jonathan Mawdsley); The Ohio State University C. A. Triplehorn Insect Collection (Luciana Musetti); University of Kansas Natural History Museum (Jennifer Thomas); Steve Bentsen; Virginia and Charlie Brown; Bob Cammarata; Steve Collins; Stephen Cresswell; Greg Lasley; Ellis Laudermilk; Larry Rosche; Judy Semroc; and Ken Tennessen.

Ashley Peters created the map of the physiographic regions of the southeastern United States. John Abbott, Wayne Barger, Daniel Dye, Mary Jane Krotzer, Ellis Laudermilk, and Mike Thomas graciously allowed their excellent photographs to be included in the book. Ed Lam created the lovely drawing of a larva in its burrow, Dr. Harold Willis graciously allowed us to use his drawing of comparison features for Margined Tiger Beetles (*Ellipsoptera marginata*) and Coastal Tiger Beetles (*E. hamata* sspp.), and Randy Beaton created the excellent tiger beetle features diagram.

Brian D. Holt would like to thank his parents and wife, Nicole. If not for their love, patience, and encouragement, he would not have the appreciation for the natural world that he does today.

Steve Krotzer would like to thank his parents for being supportive of his love of nature and his career choices, even if they didn't always understand them; his professors Bob Stiles and Mike Howell, for providing the guidance that allowed his passion for documenting the natural world to flourish; and especially Mary Jane Krotzer, for her unwavering support and unparalleled field companionship.

Giff Beaton would like to thank the following people for helping with multiple taxonomic or identification issues, some throughout all of the last 10 years we have been working on this project: David Brzoska, Skip Choate, Dan Duran, Terry Erwin, Barry Knisley, and especially Ron Huber. Dirk Stevenson has been an enthusiastic field companion and sometime tiger beetle wrangler. Mike Thomas is a great friend, sounding board, and wonderful field and photographing companion also; and I have enjoyed all of our many trips together, several of which resulted in images in this book. Finally and most importantly, I thank my wonderful wife, Allie, for her endless patience and support throughout this project and many others, and for being the best field companion ever while on many of the excursions throughout the southeastern United States that ultimately led to this book.

TIGER BEETLES

of the **SOUTHEASTERN UNITED STATES**

Introduction

Physical Description

Although they can be brightly colored and variably marked, different species of tiger beetles look remarkably similar in overall body shape and appearance. Sometimes it is necessary to look at key morphological features to distinguish similar species. A general overview and knowledge of key characteristics will help the beginning tiger beetle enthusiast to confidently identify those species they may encounter. For those interested in learning more on the subject, there are numerous texts available (Snodgrass 1935; Pearson and Vogler 2001).

Tiger beetles share the same overall body plan as other insects, having three main body segments, six legs, (usually) four wings, and

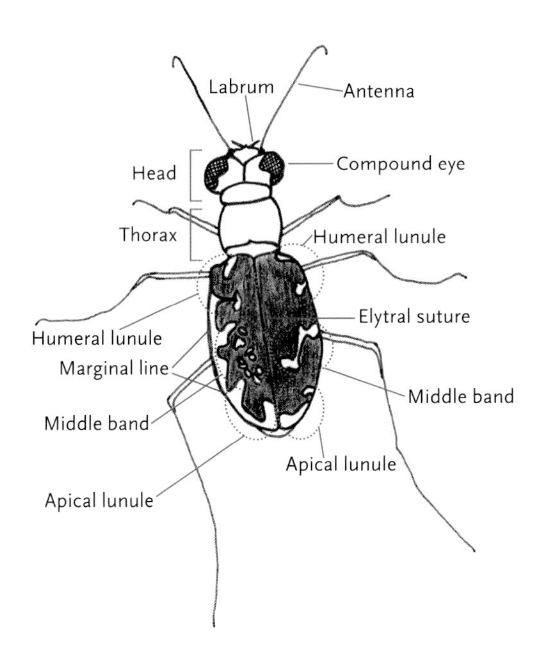

Tiger beetle anatomy. The stylized maculations are for illustrative purposes. Maculations on the left are those of *Ellipsoptera* sp. Maculations on the right are those of *Cicindela* sp. The abdomen, not labeled, lies beneath the elytra in this drawing.

two antennae. Of the three body segments, the head is the most easily recognized with its large, laterally set compound eyes. Inserted beneath the eyes on the front of the head are the eleven-segmented antennae. These function primarily as tactile sensory organs and are long and threadlike. Covering the oral cavity and many of the smaller structures constituting the mouth is the labrum, a moveable plate often referred to as the beetle's "upper lip." Labrum shape, color, and size as well as the number of teeth along the anterior edge can be useful characteristics when identifying tiger beetles. The long, sickle-shaped mandibles may be the most spectacular feature of the head. These serve many functions including capturing and manipulating prey, grasping females during mating, and defense.

The next main body segment is the thorax. The thorax consists of the prothorax, mesothorax, and metathorax. The most obvious of these is the prothorax, which is the visible "neck" area just behind the head. The upper surface of the prothorax is the pronotum, and its overall shape is useful in grouping closely related species. Attached to the thorax are the elytra. The elytra are the sclerotized, or hardened, forewings covering the membranous flight wings and the dorsal surfaces of the mesothorax, metathorax, and abdomen. In fact, the name Coleoptera, the order to which tiger beetles belong, means "sheath wing," in reference to the protective function elytra offer the flight wings.

The elytra may be immaculate (no markings) or variously marked, depending on the species. These markings, called maculations, are

Face of the Coastal Tiger Beetle (*Ellipsoptera hamata lacerata*). Baldwin County, Alabama.

the easiest way to identify most tiger beetle species in the southeastern United States.

When present, the markings occur at three areas of each elytron and are given specific names. Those located at the anterior "shoulder" area of the elytra are referred to as the humeral lunules. Midway down the elytra are the middle bands, and those occurring at the posterior are the apical lunules. The middle bands may extend partially or completely along the outer edge of the elytra so that they join the humeral and apical lunules. When this occurs, the maculation is referred to as the outer or marginal line. In certain species, the maculations may be reduced with only remnants of the lunules and middle bands occurring as spots, or they may be so greatly expanded that the entire surface of the elytra appears white. Also present on the elytra of some species are large punctures running parallel to the sutural line, the midline where the left and right elytra join.

The long, thin legs are also attached to the thorax. The basal segment closest to the thorax is the femur. Attached to its distal end is the tibia. The tibia of the front legs has specialized grooves or spines to facilitate cleaning of the antennae. The last five leg segments are collectively known as the tarsus. Males of any tiger beetle species can be determined by the presence of tarsal pads of dense setae that aid in grasping the female during mating.

The final major body segment is the abdomen. Comprising six individual segments, it is almost completely covered dorsally by the elytra. The abdomen color can be black, metallic blue, metallic green, or red, and is diagnostic for certain species. The red abdominal color

Left to right:

Six-spotted Tiger Beetle (*Cicindela sexguttata*) in flight. Jackson County, Alabama.

Typical elytral markings of the Bronzed Tiger Beetle (*Cicindela repanda repanda*). Perry County, Alabama.

flashed during flight by some species may be a warning to potential predators of the beetle's unpalatability (Pearson and Vogler 2001).

Covering the body to various degrees are setae. Simply stated, setae are the hairlike structures often found on invertebrates. These may be stiff and erect, or flattened against the body. Location and number of setae can be useful characters in tiger beetle identification, particularly when they occur along the sides of the thorax and on the labrum. However, other identification characters should also be used as these areas may become worn and bare on older tiger beetles.

The variety of colors exhibited by tiger beetles is one of the more fascinating things about the group. These colors are a result of two primary features of the tiger beetle cuticle—the presence of melanin and cuticular layering. Melanin is a black pigment occurring within the cuticle. When present in thick, disorganized deposits, light is absorbed and the beetle appears black. Where melanin is absent, light is scattered and the corresponding integument appears white or pale yellowish. These pale areas are generally restricted to the maculations, labrum, and legs of certain species. When melanin is present in organized layers, other colors are expressed dependent on the cuticular layering. This layering of melanin and wax as well as the thickness of the layers is responsible for different degrees of refraction and, as a result, different colors. Thinner layers produce violet, blue, and green colors while thicker layers produce orange, red, and magenta. When these layers are uniform, a purer color is expressed. The brilliance of many species is related to the many small pits present on the cuticular surface. Shallow pits result in a smoother surface and a more brilliant and shinier beetle (Knisley and Schultz 1997; Pearson and Vogler 2001).

Tiger beetle larva. Perry County, Alabama.

Over time the cuticle surface may become damaged. This is especially true of long-lived species of sandy substrates. The sand can eventually abrade the cuticle surface, causing a once bright and shiny tiger beetle to appear dull and worn (Knisley and Schultz 1997).

Tiger beetles are holometabolous insects; they go through a complete metamorphosis and look dramatically different as larvae than they do as adults. Overall, larval tiger beetles are soft-bodied and grublike with only the head and pronotum being heavily sclerotized. The head has four simple eyes capable of detailed focusing and three-dimensional perception, and the antennae are short and four-segmented. The prothorax, mesothorax, and metathorax are easily identified with each having a pair of well-developed legs. The ten-segmented abdomen has weakly sclerotized areas along the sides called sclerites. On the fifth abdominal segment there is a hump armed with two pairs of hooks located on the dorsal surface. This feature allows the larva to anchor itself within its burrow.

Life Cycle

Because tiger beetle larvae are largely sedentary, it is critical that the adult female choose an appropriate site for oviposition. Different tiger beetle species prefer different soil types and/or textures; other soil requirements, such as moisture and temperature, also differ among species. Each female must be able to locate just the right combination of these factors; it is thought that this is accomplished by gathering tactile information through the antennae and mandibles. The potential oviposition site also needs to be in an area that has a suitable prey base for the developing larvae, but without too many other larvae present to compete for resources.

Once a site is chosen, the female tiger beetle digs a hole in the soil with her ovipositor—interestingly, there is some evidence that the size of the ovipositor might be correlated to the preferred soil type for each species.

The female lays a single adhesive egg at the bottom of the completed hole, which may vary in depth from just a few millimeters to over a centimeter (depending on the species); she then "backfills" the hole and covers the entrance to make it less obvious to potential egg predators.

Relatively little is known about the egg itself, but not surprisingly, there is a correlation between egg size and beetle size; in other words,

Left to right:

Female Northern Barrens Tiger Beetle (*Cicindela patruela patruela*) ovipositing. Wolfe County, Kentucky.

Highlands Tiger Beetle (*Cicindelidia highlandensis*) covering oviposition site. Polk County, Florida.

big tiger beetles lay big eggs and small tiger beetles lay small eggs. Regardless of size, the eggs take about two weeks to complete development; at this point, if conditions such as soil moisture are favorable, the egg will hatch.

Upon emergence from the egg, the larval exoskeleton is soft and delicate for the first day or so. Once the exoskeleton has fully hardened, the larva begins to excavate a burrow from the hole in which the egg was laid. The larva uses its mandibles and dorsal hooks to loosen dirt from the sides of the hole. It carries the dirt to the surface on its flat head and flips small pellets of dirt backward from the entrance.

Tiger beetle burrows differ from those of most other burrowing invertebrates in that the soil pellets are almost always cast in the same general direction, so instead of a ring of pellets around the hole, there is a pile just on one side. The process is repeated until the burrow is large enough to accommodate the larva yet tight enough for the larva to easily wedge itself against the sides of the burrow at the entrance to provide leverage to pull prey down and to protect against being pulled out by a predator. The burrow is also excavated to a depth that provides the developing larva with protection from predators and a shelter from temperature extremes.

Tiger beetle larvae go through three stages called instars or stadia: in general, the first instar lasts only a few weeks; the second will last a month or two; and the final instar can last from a few months up to a year or longer. As the larva transitions from one instar to the next, it sheds its old exoskeleton by molting to allow for growth; during the

molting process the larva is inactive and will be helpless for several days, so it plugs the burrow entrance for protection. After each molt the tiger beetle larva must enlarge and deepen its burrow to accommodate its growing body. By the time a larva reaches its final instar, the burrow can reach 30 inches in depth or more, depending on the species. Burrows are usually more or less perpendicular to the soil surface, but they can sometimes be constructed at an angle.

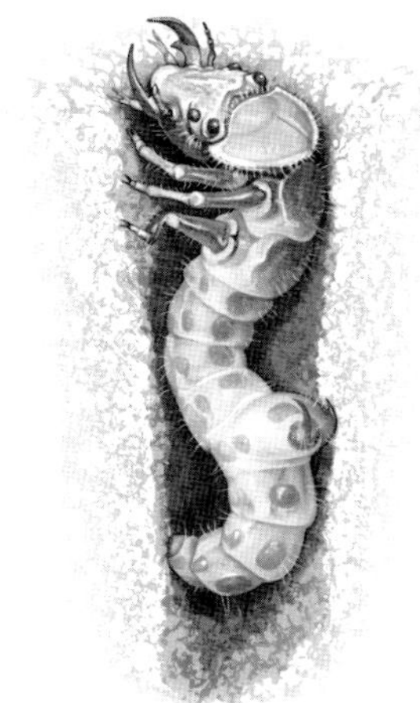

Top to bottom:

Tiger beetle larva excavating burrow. Liberty County, Georgia.

Drawing of tiger beetle larva and burrow.

The burrow mouth is a simple round opening, but some species will also build funnels and pits around the mouth to help in capturing prey; other species incorporate a shady overhang to the entrance, presumably to help maintain the tunnel at a suitable temperature or moisture level.

Tiger beetle larvae can take from one to four years to complete their development, depending on the species and on food availability. Each successive instar generally takes longer to complete than the previous one, and as larvae grow larger, they require more food to maintain themselves. The final instar, in particular, requires sufficient food in order for the larva to store energy for metamorphosis. When larval development is complete, the burrow entrance is sealed, and the larva retreats to the bottom of the burrow and builds a small pupal chamber. Although relatively little is known about the specifics

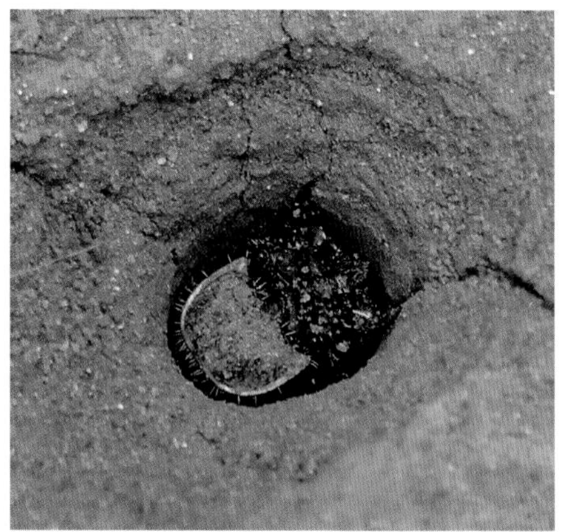

Left to right:

Cicindela larva in feeding position at top of burrow. Glynn County, Georgia.

Tetracha larva in feeding position at top of burrow. Carroll County, Georgia.

of tiger beetle metamorphosis, it is believed that the pupal stage lasts for a few weeks in most species. During this time, the adult structures are developing within the pupal skin.

Another process for which some details remain unknown in tiger beetles is ecdysis, the final molt in which the adult tiger beetle emerges from the pupal case. The emerging adult, through a combination of air swallowing, blood pumping, and rhythmic movements, exerts outward pressure on the pupal skin, eventually causing it to split in several places. The adult beetle then pushes/pulls itself free from the remnants of the pupal case. Upon emergence the adult is pale, soft, and defenseless, so it remains in the safety of the burrow for several days while the exoskeleton hardens. Adult coloration also develops during this time. Then the adult exits the burrow by digging through the plugged entrance hole or, less often, by digging an entirely new exit tunnel. Adults of some species leave the larval burrow as soon as they've hardened, while others remain until some environmental cue (soil temperature, soil moisture, etc.) triggers their exit.

The process of metamorphosis is complex, and every individual tiger beetle doesn't survive to adulthood. Even those that do are sometimes malformed in some way: they may be missing part or all of a leg; the mandibles may be misshapen; the elytra may be fused together. These abnormalities most likely occur during ecdysis, and they may significantly impact the individual's ability to capture prey or escape from predators. Although such individuals are seen occasionally, it is not clear whether they live long enough in the wild to reproduce.

Most tiger beetles in the southeastern United States have a one- or two-year life cycle. The proportion of this time spent from the egg to pupal stage varies somewhat depending on food availability, weather, and other variables, but this period definitely makes up the majority of the life cycle. Adult tiger beetles, at least in the Southeast, typically live only for a few weeks, perhaps a couple of months at most.

Mating and Mating Behaviors

Most tiger beetles become reproductively active shortly after emergence. There is some evidence that males, at least of some species, may emerge prior to females; this adaptation would presumably give these males an advantage in finding a female, and especially in finding an unmated female as she emerges from the pupal burrow. In contrast with many other insect groups in the region, male tiger beetles do not display any territorial or courtship behaviors and are characterized as "opportunistic" breeders. When a reproductively active male tiger beetle encounters a female, he will chase her down much in the same way that he would capture prey. Grasping the female by the thorax with his mandibles, which fit into grooves on the female's thorax for a better grip, he will climb atop her. Males also use pads of modified setae present on the front legs to help maintain their grip. Many females will actively and repeatedly try to dislodge the male, perhaps testing his potential as a mate or signaling her unwillingness to mate. The males are persistent, though, and more often than not are able to maintain their position. At this time copulation usually occurs, and a specialized capsule of sperm cells called a spermatophore is transferred from the male to the female.

There is some evidence that, as in some other groups of insects, male tiger beetles in some species have the ability to remove spermatophores stored by the female from previous matings. When not actually copulating, males will often remain firmly attached to the female; this behavior is called contact guarding and is believed to be a mechanism to prevent other males from mating with this female. The combination of the ability to remove competing males' spermatophores and contact guarding to prevent his spermatophore from subsequently being removed is an important adaptation to maximize the chances that a particular male can pass his genetic material along to the next generation of tiger beetles.

Natural hybridization in tiger beetles is apparently rare, which might seem surprising given that the males are often very indiscrim-

Left to right:

Coppery Tiger Beetles (*Ellipsoptera cuprascens*) mating. Hale County, Alabama.

A male Moustached Tiger Beetle (*Ellipsoptera hirtilabris*) contact guarding a female. Lowndes County, Georgia.

inate in their mate choice and will readily attempt to mate with females (or males!) of other species. It is believed that the male mandibles and the grooves on the thorax of the female have coevolved into a "lock and key" mechanism that is species specific. When a male establishes contact with a female of a different species, she will be able to sense that something is not right, and she will then reject any mating attempt and will eventually be able to dislodge the male. In those cases where the male is able to maintain his grip and attempt to copulate, the male and female genitalia are also believed by some scientists to provide a second level of "lock and key" protection; if the two sets of genitalia don't fit together properly, then the transfer of the spermatophore will be unsuccessful. This double layer of protection isn't foolproof, however; hybrids between some species can be easily produced in the laboratory, and occasionally individuals are found in the wild that apparently represent hybrid offspring of closely related, co-occurring species. It is not known at this time whether these apparent hybrids are capable of reproducing.

Much of the above information has been gleaned from casual observation and other anecdotal evidence. Research into the reproductive behavior of tiger beetles has been relatively limited to date, especially in the southeastern United States, and there is much left to learn about this aspect of these fascinating insects.

Feeding Behavior and Prey

Adult tiger beetles are highly active visual predators, preying on a variety of smaller arthropods. Ants are probably the most frequently observed prey due to their common co-occurrence with tiger beetles; however, certain species of ants are avoided by tiger beetles, presumably due to some defensive chemical emitted by the ants.

Other ant species are capable of attacking or harassing adult tiger beetles until they are driven away. There is some evidence that adult tiger beetles actually prefer soft-bodied prey. The authors have observed instances where adult tiger beetles ignored numerous ants to pursue ladybug larvae and other softer prey, such as tridactylid cricket nymphs; even caterpillars many times larger than the tiger beetle can be subdued.

The fast, agile adults usually hunt visually, capturing moving prey by running toward it and pausing occasionally to adjust their approach in relation to the prey's current location. Ambush hunting, although observed less often, can be another effective predation strategy for tiger beetles. Some species, particularly those active at dusk or at night, are possibly able to locate their prey by sound (Guido and Fowler 1988). Adult tiger beetles have been known to hunt on water, using vegetation such as lily pads to travel over the surface of ponds.

If a captured potential prey organism is too large to subdue or emits an offensive chemical, the adult tiger beetle will generally release it quickly. Otherwise, the tiger beetle uses its mandibles to pierce and cut the food item into smaller pieces and its mandibular palpi to help work the macerated food into a ball called a bolus. Digestive enzymes are secreted on the bolus, and the resulting fluids are transferred to the mouth. Any heavily sclerotized, indigestible body parts are discarded by the tiger beetle without being eaten.

Adult tiger beetles are opportunistic feeders and will occasionally scavenge for food; documented food items include dead insects, carrion of various vertebrates including fishes (Grammer 2009), stranded tadpoles, and even fallen fruit. These food items are generally located by touch, utilizing the tiger beetles' sensitive antennae and palpi.

In areas with multiple species of tiger beetles in one location, the amount of food available at any one time can be a limiting factor. Adults will often segregate themselves to avoid, or at least minimize, competition for food. In some cases the segregation may be spatial,

Left to right:

Coppery Tiger Beetle (*Ellipsoptera cuprascens*) feeding on a winged ant. Sumter County, Alabama.

Bronzed Tiger Beetle (*Cicindela repanda repanda*) with tridactylid prey. Grady County, Georgia.

with different species occupying slightly different parts of a habitat. For instance, along a river shoreline bar, one species might occupy the fine damp sand right along the water's edge; a second species the slightly higher, coarser, and drier sand a bit farther from the water; a third species the dry cobble area at the upstream end of the bar; a fourth species the high, dry white sand at the upper edge of the flood zone; and a fifth species the moist, shaded soil where the forest meets the bar. In this way, multiple species can successfully occupy one high-quality habitat. A second type of segregation may be temporal, where different tiger beetle species are active at different times of the day, year, or both. For example, one species might be most active for the first few hours after sunrise during the spring and fall, while another might be most active during the middle of the day during the summer, and a third species might be nocturnal and active from early spring through late fall. And in instances where segregation does not appear to be occurring and multiple species of tiger beetles are using the same habitat, there is some evidence that the effects of competition could be mitigated by slight differences in preferred prey size among the species.

Tiger beetle larvae, like the adults, feed on a wide variety of smaller insects, crustaceans, spiders, and the like. Late instar larvae of larger species are capable of capturing quite large prey, including grass-

Top to bottom:

Bronzed Tiger Beetles (*Cicindela repanda repanda*) scavenging a rodent carcass. Perry County, Alabama.

Carolina Metallic Tiger Beetle (*Tetracha carolina carolina*) scavenging a tree frog carcass. Baker County, Florida.

hoppers, dragonflies, and even adult tiger beetles. They are classic ambush hunters, waiting in place at the top of the larval burrow for unsuspecting prey to wander by. The larval head and pronotum fill the entrance, and the legs, hooks, and setae hold the larva in position. When prey is detected, either visually or by sensing the tiny vibrations it causes, the larva strikes incredibly quickly, faster than can be seen by the naked eye. Rearing up and usually backward out of the burrow, up to as far as half its body length, the tiger beetle larva grabs the hapless victim with powerful mandibles and pulls it back into the burrow. Occasionally an exceptionally large prey organism will be too large to fit into the burrow, and the tiger beetle larvae will remain at the burrow mouth with its victim. As with the adults, actual feeding is accomplished by macerating the prey into a soft bolus, predigesting with enzymatic fluids, and consuming the resulting liquefied tissues. Indigestible remains are usually carried back up to the top of the burrow and thrown away from the entrance.

Tiger beetle larvae have to eat a certain amount of food (called the "biomass threshold") before they can molt to the next instar; if they cannot meet this threshold, a few will adapt and molt at a smaller size, but most will die. The ability to consume enough food to meet this threshold at each successive molt not only influences the size of the larvae, but it also has an impact on the size of the adult and its ability to produce a normal amount of offspring.

Predators and Parasitoids

Adults

Adult tiger beetles are very skilled at avoiding humans wielding either a net or a camera, but there are more formidable predators from which they must also escape. Many different animals will prey on tiger beetles from time to time if the opportunity arises. Groups of hunting ants are capable of attacking and subduing an adult tiger beetle; ant heads found on the legs or antennae of living tiger beetles are thought to be evidence of an unsuccessful attack. Orb-weaving spiders will occasionally catch a careless tiger beetle in flight; jumping and wolf spiders ambush, kill, and consume the adults. Frogs and toads are also occasional ambush predators of adult tiger beetles. Based largely on evidence provided by the examination of their scat, it appears that several small mammals such as armadillos, skunks, and raccoons include tiger beetles in their diet.

The more important predators of adult tiger beetles are those that specifically target and hunt them, such as some insectivorous birds, lizards, and flies in the family Asilidae, also known as robber flies. Most birds that pursue adult tiger beetles (such as blue jays and various flycatchers) capture them during attempted escape flights; other birds such as shrikes, however, are more likely to capture tiger beetles on the ground. It has been observed that birds will sometimes alter their feeding behavior to take advantage of those times when adult tiger beetles are especially abundant. Plovers and other sand-inhabiting shorebirds have been observed gorging on tiger beetles to feed their chicks. Lizards also are adept at capturing adult tiger beetles, especially when temperatures are relatively cool and the insects cannot effectively fly to escape.

Perhaps the most efficient tiger beetle hunters are robber flies. These aggressive predatory flies hunt from a perch, either on or above the ground. When a potential prey organism (almost always flying insects smaller than themselves) is spotted, the robber fly pursues its target, captures it in flight, injects it with a powerful paralyzing toxin mixed with strong digestive enzymes, and carries it back to the perch. There it feeds on the partially digested prey with specialized mouthparts. Some of the larger robber flies in the Southeast occur in areas where tiger beetles are common, and they have the potential to be significant predators. There is some anecdotal evidence that some robber flies might follow large animals, including humans, capturing insects that are flushed into flight. One of the authors was part of such an event with a large western species of robber fly and a slightly smaller species of tiger beetle. In the course of about 10 minutes during a photography excursion, three different individual tiger beetles, flushed into an escape flight by the activity of the photographer, were seized and eaten by the robber fly (much to the displeasure of the photographer and, one assumes, the tiger beetles!).

Larvae

Larval tiger beetles spend most of their lives hidden in underground burrows, so it is not surprising that there have been very limited observations of larval predation. Ants will occasionally kill and consume a larva, and some birds, such as flickers, are capable of removing the larva from its burrow. Far more significant enemies of the larval tiger beetle, however, are a few species of parasitoid wasps and flies. These insects lay eggs on or near the tiger beetle larva; upon hatching, the developing wasp larvae or fly maggots feed on the tiger beetle, eventually killing it (in contrast with a parasite, which does not kill its host). Parasitoid wasps and flies use strikingly different methods. The female wasps—tiny, ant-like, often wingless members of the family Tiphiidae—find the tiger beetle larva in its tunnel. They must be able to get close enough to sting and paralyze the tiger beetle; some species accomplish this by masquerading as hapless prey, allowing the

Tiger beetle larva parasitized by a wasp larva (*Methoca* sp.). Perry County, Alabama.

tiger beetle to seize them. The larva attempts to subdue the wasp but instead is stung and paralyzed. The wasp lays a single egg on the paralyzed larva and leaves, plugging the entrance to the tunnel to protect its investment. The wasp egg hatches in 4 or 5 days; the developing wasp larva consumes the tiger beetle larva, then pupates in the burrow before emerging as an adult. Female bombylid flies in the genus *Anthrax*, on the other hand, locate a tiger beetle larval burrow, hover over the burrow opening, and flip eggs into the tunnel entrance. The eggs roll to the bottom of the burrow, hatch, and eventually the tiny fly maggot will find and attach to the tiger beetle larva. The maggot remains there until the tiger beetle is ready to pupate; at this point the fly maggot speeds up its development, consuming the tiger beetle host in the process. A tiger beetle larva found out of its burrow has often been parasitized in this manner.

These parasitoid wasps and flies occur and are common throughout the southeastern United States, especially during the summer; it is likely that they have a limiting effect on the success of tiger beetle larvae of several species in the region.

In addition to the predators and parasitoids mentioned above, tiny mites are often found attached to both adult and larval tiger beetles. While it is certainly possible that these mites are parasites, some researchers feel that they are phoretic (just "hitching a ride").

Defensive Strategy

Not surprisingly, tiger beetles have evolved multiple defense mechanisms to help outwit these predators and parasitoids. Adult tiger beetles have exceptional vision, are fast and agile runners, and in many species are quick to take to the air in an escape flight. If this rather straightforward method of detection and escape doesn't work, tiger beetle adults have several other defensive strategies at their disposal. Perhaps the simplest method of predator avoidance is hiding; most adult tiger beetles will shelter in burrows, vegetation, or under debris when not active. They will often dig temporary burrows to escape threats or wedge into cracks and crevices in the ground.

Several species are crepuscular or nocturnal; in this way, they largely avoid diurnal predators such as robber flies. A very common defense strategy is camouflage, and many tiger beetles are colored and/or patterned to blend nearly perfectly with the substrate on which they occur. For instance, the mostly white Ghost Tiger Beetle (*Ellipsoptera lepida*) can virtually disappear against the light-colored dry sands of its typical habitat. Some tiger beetles exhibit disruptive coloration, a different type of camouflage in which a pattern of strongly contrasting elytral maculations serves to break up their outline, making it difficult for predators to recognize the beetle as potential prey.

Still other tiger beetles are colored to match objects in their habitat, rather than the substrate. The small Eastern Pinebarrens Tiger Beetle (*Cicindelidia abdominalis*) is strikingly similar in size to small dark objects on the white sands where it is found. This similarity apparently helps to keep potential predators from recognizing the black beetle as a prey, since they can surely see it against the white sand. Lastly, some tiger beetles that are very conspicuous when at rest actually become cryptic in flight. The bright green Six-spotted Tiger Beetle (*Cicindela sexguttata*) is quite easy to follow as it forages on the ground along forested openings. However, when flushed into flight, it becomes almost impossible to see against the sun-dappled green vegetation in its habitat.

Left to right:

Burrow excavation by Ghost Tiger Beetles (*Ellipsoptera lepida*). Perry County, Alabama.

Disruptive camouflage of the Whitish Tiger Beetle (*Ellipsoptera gratiosa*). Lexington County, South Carolina.

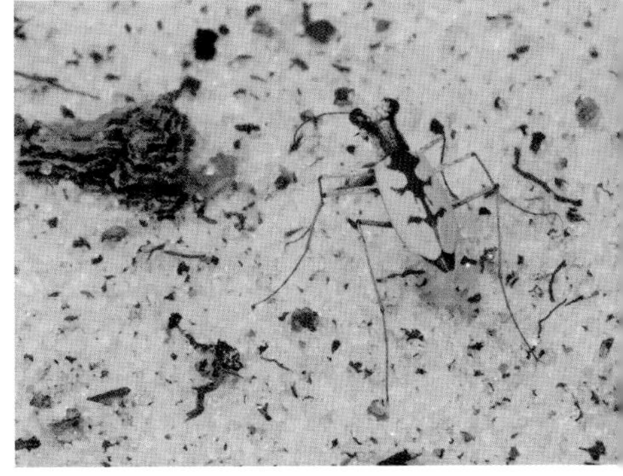

In contrast to species that depend on concealment for survival, some adult tiger beetles have bright or bold colors that are thought to serve as aposematic coloration—a warning to potential predators that the beetles are poisonous, or at least very distasteful. Several species of American Tiger Beetles (*Cicindelidia* spp.) have reddish-orange abdomens that are conspicuous in flight and can sometimes be seen when the beetle is at rest; these same species tend to possess noxious chemicals, such as benzaldehyde and cyanide, that can be released upon capture. It is likely that the brightly colored abdomen in these species has evolved as a warning mechanism to deter certain predators.

Tiger beetle larvae have relatively few antipredator defense mechanisms; they depend mostly on a quick retreat down the burrow when threatened. Recently, though, an unusual method of larval wheel locomotion in one species has been discovered. When threatened, the larvae of the Eastern Beach Tiger Beetle (*Habroscelimorpha dorsalis* sspp.) are able to form a continuous loop by grasping the tip of their abdomen with their mandibles; they then launch themselves in the air and escape by allowing the prevalent beach winds to blow them along the sand, much like a child's hoop toy (Harvey and Zukoff 2011). Tiger beetle larvae also have excellent vision, perhaps even better than that of the adults, and can see potential predators that approach the burrow opening. It is thought by some that the larvae are able to sense the vibrations of potential predators in the vicinity of the burrow; the senior author has witnessed this ability on numerous occasions when attempting to photograph larvae at the mouth of their burrow. Finally, the hooks that help secure the larvae in the burrow also make it more difficult for a predator to extract them.

Seasonal Patterns

All tiger beetles in the southeastern United States basically can be categorized as either summer or spring/fall species, depending on the timing of the adult emergence from the pupal stage. In nearly all summer species in the southeastern United States, the life cycle is completed in one year. Adults emerge from the pupal burrows during early summer, and they complete their life cycle (mature, reproduce, and die) during a relatively brief period. The developing larval offspring of these adults spend the following winter and early spring in their underground burrows before pupating and emerging the following summer. In a very few species, such as the Ghost Tiger Beetle (*Ellipsoptera lepida*), a summer species takes two years to complete its life cycle; in this case, early instar larvae overwinter during the first winter, complete larval development during the first summer, late instar larvae overwinter in the second winter, and adults emerge during the second summer (Shelford 1908). In our mild climate, many summer species persist longer than might be expected; the nocturnal Metallic Tiger Beetles (*Tetracha* spp.) have a particularly extended period of adult activity, often beginning fairly early in the spring and lasting well into the fall.

The second emergence pattern is exhibited by those tiger beetles known as spring/fall species (or, perhaps more accurately, fall/spring species, as some experts have begun to refer to them). In these species, transformation from the pupal stage to adulthood takes place in the late summer or fall; adults are active for several weeks but are not reproductively mature and do not mate at this time. With the onset of colder weather, the adults return underground to overwinter. The following spring the adults reemerge for a second time and complete their life cycle (mate, lay eggs, and die). Some species can be quite long-lived as adults, especially in the South; the Bronzed Tiger Beetle (*Cicindela repanda*) is the best example. A few adults from the spring cohort will still be active when the fall cohort emerges, resulting in a combination of very old and very "fresh" individuals in the same population. The entire spring/fall life cycle takes two to three years to complete, depending on the species and the local climate conditions; most species in the Southeast have a two-year cycle. Those species with a two-year life cycle overwinter as adults; those species with a three-year cycle overwinter as larvae in the first year and as adults in the second.

There are a few tiger beetle species that have a modified version of the typical spring/fall emergence pattern. The most notable is the Autumn Tiger Beetle (*Cicindela nigrior*), which completes its entire adult cycle during the fall of the year and is not present at all in spring; it is the only spring/fall species in the region that normally reproduces in the fall. Another very common species, the Six-spotted Tiger Beetle (*C. sexguttata*), lacks the fall emergence period, instead emerging in the spring and remaining active for most of the summer.

In general, members of the genera *Apterodela*, *Cicindelidia*, *Ellipsoptera*, *Eunota*, *Habroscelimorpha*, and *Parvindela* are summer species, at least in the Southeast. Most southeastern coastal and riverine tiger beetles are summer species, but this group also includes several upland species. Most of the spring/fall species are in the genus *Cicindela*, and many are upland species that reach the southern limit of their overall distribution in the southeastern United States. It is not unusual for a given area to have a combination of tiger beetle species representing both types of life cycle patterns; it is likely that their differing emergence times allows for more efficient utilization of resources such as food and shelter by reducing competition for these resources.

Habitats

With few exceptions, visual predators like tiger beetles—that use their superior eyesight to acquire prey and their superior speed to catch it—need open spaces to see their prey and run it down. For these reasons, most tiger beetle habitats are open: dry sand, bare dirt, river and coastal shorelines, or unvegetated trails. This section groups species by habitat. Due to sometimes minor differences between habitat requirements, some species are more widely found than others, but these groupings should help readers recognize appropriate habitats in which to search for tiger beetles and for which species to look. There are exceptions to these groupings as this section is too general to list every detail for every species, but additional pertinent details can be found in each individual species account.

Broadly speaking, all habitats can be grouped by the "ecoregion" in which they occur. Important characteristics of each physiographic or ecoregion include elevation, soil type, rainfall, geology, and other factors, and there are many ways to divide each region into smaller subregions. Here the southeastern United States is broken down into six major ecoregions: Appalachian Plateau, Blue Ridge, Coastal Plain, Interior Low Plateau, Piedmont, and Ridge and Valley. From highest in elevation to lowest, a brief description of each ecoregion is included here.

The Blue Ridge contains all of the Southeast's highest mountains and has the most northern fauna, including many species found nowhere else in the region (it is also a region of high biodiversity of both flora and fauna). Elevations range from 1,600 to almost 6,700 feet (Mount Mitchell in North Carolina is the highest point east of the Mississippi River), and it is a highly eroded ancient mountain system composed primarily of metamorphic rock. The Blue Ridge as a whole is the coldest and rainiest region in the Southeast.

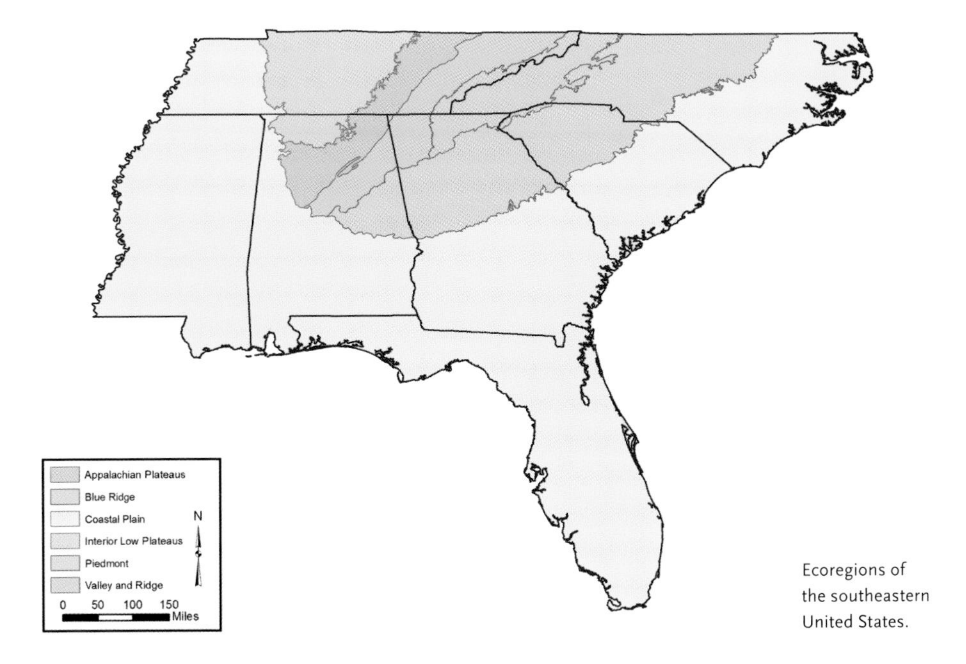

Ecoregions of the southeastern United States.

The Appalachian Plateau (sometimes called the Cumberland Plateau) is a high, flat plateau that runs from north-central Alabama through the extreme northwestern tip of Georgia and up through central Tennessee. It is made up primarily of sedimentary rocks and forms the western edge of the Appalachian Mountains.

North and west of the Appalachian Plateau is the Interior Low Plateau, which comprises much of middle Tennessee and the extreme north-central part of Alabama. The terrain is mostly unglaciated limestone loosely organized into a mixture of rolling plains with large rivers and some steeper topography along the plateau rims. The presence of numerous large caves in this region is a reflection of its limestone sediments and karst topography.

Just east and south of the Appalachian Plateau is the Ridge and Valley, which parallels the plateau. This region includes a sizeable portion of northeast Alabama as well as much of northwestern Georgia and eastern Tennessee. As the name implies, this is an area of alternating narrow ridges and wide valleys.

The Piedmont is the next ecoregion to the south and east, as well as the next lowest in elevation. It stretches from central Alabama in the west through north-central Georgia and constitutes much of the western halves of both Carolinas. The southeastern border of the Piedmont is a unique geological feature known as the Fall Line, which is a line of waterfalls and rapids where rivers pass from the higher elevations of the Piedmont into the lower, and flatter, Coastal Plain. The Piedmont is mostly characterized by rolling topography and larger rivers than the Blue Ridge and the Ridge and Valley to the north.

South of the Fall Line lies the Coastal Plain. This ecoregion encompasses the majority of the Southeast: far western Tennessee, all of Mississippi and Florida, and the southern or southeastern portions of Alabama, Georgia, South Carolina, and North Carolina. The Coastal Plain is fairly flat and is mostly sandy. Elevations are low, ranging from near 0 to 500 feet, and temperatures are warm; the Coastal Plain has even larger rivers than the Piedmont with attendant wide areas of floodplain. There are several slightly different subtypes of coastal plains in the Southeast, but for the purposes of this book we have combined them into one ecoregion.

Coastal Sandy Beach and Adjoining Marsh Edges

This habitat type includes all Atlantic Ocean/Gulf of Mexico saltwater beaches (gulf and bayside) and adjacent marsh-mud or other tidal flats except salt panne (see Coastal Plain Salt Panne). Common species on sandy beaches include all subspecies of Eastern Beach Tiger Beetles (*Habroscelimorpha dorsalis* sspp.) and, less commonly, the coastal form of the Hairy-necked Tiger Beetle (*Cicindela hirticollis*).

All beach species are more common on beaches with less foot and, especially, less vehicular traffic. All of these beetles may move up and down the beach in relation to the tide. Muddy pools on upper beaches or near marshes are the best places to look for Margined (*Ellipsoptera marginata*) and Coastal (*E. hamata* sspp.) Tiger Beetles as well as S-banded Tiger Beetles (*Cicindelidia trifasciata*), but these species may sometimes be found on nearby sand or other surfaces including boardwalks.

All three species of Metallic Tiger Beetles may also be found in or near this habitat; the Florida Metallic Tiger Beetle (*Tetracha floridana*) is largely restricted to estuarine tidal flats, but the two more widespread species—the Carolina Metallic Tiger Beetle (*T. carolina*) and Virginia Metallic Tiger Beetle (*T. virginica*)—are found throughout the region in many habitat types. All three are very strongly nocturnal and are rarely seen during the day when most tiger beetle enthusiasts are out looking.

Top to bottom:

Sandy beach habitat. Franklin County, Florida.

Muddy tidal flat habitat. Charleston County, South Carolina.

Florida Metallic Tiger Beetle (*Tetracha floridana*) habitat. Lee County, Florida.

Coastal Plain Salt Panne

This is a very specific habitat consisting of shallow pools in brackish marsh areas that go dry between tides and have higher salinity than surrounding areas, leading to the establishment of a specific suite of plants and animals. Salt pannes are uncommon in the Southeast, but occur sporadically along the coasts in the region, especially the Gulf of Mexico coast from Florida to Mississippi. While the more common species of coastal tiger beetles (Margined, Coastal, and S-banded Tiger Beetles) also occur in salt pannes, the lures for tiger beetle enthusiasts are the far more restricted species that are pretty much found

Top to bottom:

Coastal salt panne habitat. Dixie County, Florida.

Elusive Tiger Beetle (*Eunota striga*) habitat. Hillsborough County, Florida.

only at salt pannes in the region. The three species to look for (within their mapped ranges) are the Saltmarsh (*Eunota severa*), Gulfshore (*E. pamphila*), and White-cloaked Tiger Beetles (*E. togata*).

One additional species sometimes found here is the aptly named Elusive Tiger Beetle (*E. striga*), which can be found near salt pannes or in other brackish grassy areas away from pannes; however, it is not usually found out in the open like the other species in this section.

Coastal Plain Sandy Rivers and Streams

This habitat type is very common in the region and occurs along the edges of all streams and rivers in the Coastal Plain, which are typified by sandy shorelines and bars. The Bronzed Tiger Beetle (*Cicindela repanda*) is abundant throughout the region (except peninsular Florida) in this habitat, but some of the other species that occur here are more selective in their requirements. Just like at saltwater beaches, the S-banded Tiger Beetle usually needs a muddy component to the sand, as does the inland Hairy-necked Tiger Beetle. Both are fairly uncommon along rivers, although the closer one is to the coast, the more common S-banded becomes. Also found on sandy beaches that sometimes have a mud component, but restricted to the western part of the region, are the lookalikes Coppery Tiger Beetle (*Ellipsoptera cuprascens*) and Sandy Stream Tiger Beetle (*E. macra*), as well as the Ocellated Tiger Beetle (*Cicindelidia ocellata*).

Mud/sand river habitat. Tattnall County, Georgia.

Top to bottom:

White sand river habitat. Stone County, Mississippi.

Cobblestone Tiger Beetle (*Cicindelidia marginipennis*) habitat. Perry County, Alabama.

The next four species are restricted to streams and rivers with drier, whiter sand. The Sandbar Tiger Beetle (*Ellipsoptera blanda*) and White Sand Tiger Beetle (*E. wapleri*) are often found together near the water in the appropriate range; Big Sand Tiger Beetles (*Cicindela formosa*) and Ghost Tiger Beetles (*E. lepida*) are usually found in drier areas of white sand (and sometimes either species can be found away from water if there is enough sand).

Included in this category because there is nowhere else to put it is the Cobblestone Tiger Beetle (*Cicindelidia marginipennis*), which

Ant-like Tiger Beetle (*Parvindela cursitans*) habitat. Hale County, Alabama.

requires a river bar that contains cobble in addition to gravel or sand. This habitat type is extremely rare in the Southeast, and so is this beetle.

The last beetle in this category is the Ant-like Tiger Beetle (*Parvindela cursitans*), which has even more specific habitat requirements: the loamy transition zone between riverbank forest and sandy river edge.

Coastal Plain Inland or Dry Sand

Another broad category of tiger beetle habitat in the Coastal Plain is dry sand, usually in the form of an open field or clearing or along forest trails and roadway edges. The tiger beetles discussed below occur, and sometimes co-occur, in this dry, sandy habitat. The most commonly encountered species is the Punctured Tiger Beetle (*Cicindelidia punctulata*), which can be found literally anywhere the substrate is dry and even minimally open. Two additional species found in dry, sandy habitat are the similar Festive Tiger Beetle (*Cicindela scutellaris* sspp.) and Autumn Tiger Beetle (*C. nigrior*); the Autumn Tiger Beetle prefers slightly more vegetated areas than the Festive Tiger Beetle, which prefers larger openings mostly free from vegetation.

Two smaller, very pale tiger beetles, the Whitish Tiger Beetle (*Ellipsoptera gratiosa*) and Moustached Tiger Beetle (*E. hirtilabris*), are mostly confined to very white or sugar sand habitat, where they can be difficult to spot on the pale, dry sand.

Top to bottom:

Coastal plain open sand habitat. Okaloosa County, Florida.

Coastal plain "sugar" sand habitat. Levy County, Florida.

Finally, there are four somewhat similar very small tiger beetles also found in the dry Coastal Plain habitat. The first of these is the extremely rare Miami Tiger Beetle (*Cicindelidia floridana*), confined to pine rockland habitat near Miami, Florida (there is very little of this habitat left).

Pine rockland habitat of the Miami Tiger Beetle (*Cicindelidia floridana*). Miami-Dade County, Florida.

A second beetle with a very restricted range is the Highlands Tiger Beetle (*Cicindelidia highlandensis*), found only along the Lake Wales Sand Ridge in central Florida. The other two species are more widely distributed in open or semi-open sandy habitat. The Eastern Pinebarrens Tiger Beetle (*C. abdominalis*) occurs in several states in the region, while the Scabrous Tiger Beetle (*C. scabrosa*) is found throughout most of peninsular Florida.

Upland Bare Soils

Upland bare soils are another habitat type favored by certain species of tiger beetles; these sites can occur naturally through fire or, more often, are created by human activities. We use the term "upland" here to mean anything out of the Coastal Plain. One of the ways this habitat becomes available is when a lot is cleared for construction but the construction never happens—within a couple of years this area will be colonized by tiger beetle species that occur nearby.

The Eastern Red-bellied Tiger Beetle (*Cicindelidia rufiventris* sspp.) is found throughout the upland bare soils of the region (and can

Upland bare soils habitat. Union County, Georgia.

sometimes be found in similar habitat on coastal plains). The Oblique-lined Tiger Beetle (*Cicindela tranquebarica*) also occurs most commonly in the upland bare soils habitat. The Splendid Tiger Beetle (*C. splendida*) is most commonly found in open upland bare soils in the Blue Ridge. The Cow Path Tiger Beetle (*C. purpurea*), rare in the Southeast, is also primarily confined in this region to Blue Ridge bare soils.

Upland Forest Trails

Upland forest trails are similar to the bare soil habitat type but consist of long, narrow patches of habitat as opposed to open areas. This habitat type can be very patchy as portions of trails become vegetated while others remain open.

All of the species listed in upland bare soils can also be found on forest trails, especially the Eastern Red-bellied Tiger Beetle. The Six-spotted Tiger Beetle (*Cicindela sexguttata*) is also most commonly found along forest trails. The Northern Barrens Tiger Beetle (*C. patruela*) is almost extirpated from the region, but the last few spots where it was or is known to occur in the Southeast are small forest trails at higher elevations.

One-spotted Tiger Beetles (*Apterodela unipunctata*) are distributed throughout upland forests, but the only places to find them regularly are along forest trails. They are never found in large numbers but are patchily distributed throughout the habitat.

Top to bottom:

Upland forest trails habitat. Lumpkin County, Georgia.

Northern Barrens Tiger Beetle (*Cicindela patruela patruela*) habitat. Wilkes County, North Carolina.

Upland River Edges

River and stream edges are also good tiger beetle habitats in the upland areas of the Southeast; upland streams are generally much rockier than those in the Coastal Plain, with limited amounts of sand or sandy areas. The very common Bronzed Tiger Beetle is found along

Upland river edge habitat. Haywood County, North Carolina.

almost all rivers and streams in the uplands, and the Twelve-spotted Tiger Beetle (*Cicindela duodecimguttata*) can also be found here, especially along streams with mixed rock/dirt edges. Twelve-spotted Tiger Beetles may also be found around ponds and lakes. The rare Appalachian Tiger Beetle (*C. ancocisconensis*) is found only along a few rivers in the Blue Ridge and Ridge and Valley Ecoregions.

Taxonomy

Tiger beetles were first described by Carolus Linnaeus, the "Father of Taxonomy," in the 1700s. Since then, their position within the order Coleoptera has changed. Tiger beetles have been historically placed in their own family (Cicindelidae), based on certain morphological characters of the adults and larvae: adult antennal placement narrower than clypeus (the facial plate between the labrum and antennal insertions), both spurs of adult front tibiae apical, and a hump with hooks on the fifth abdominal segment of larvae. In the last 20 years, a few molecular and phylogenetic studies suggested tiger beetles were more correctly placed within a subfamily (Cicindelinae) of the ground beetle family (Carabidae). However, the taxonomic position of tiger beetles has recently been reevaluated and determined that placement in their own family (still Cicindelidae) is justified (Duran and Gough 2020). We follow this placement for this guide.

A number of important people have contributed to the knowledge of tiger beetle classification. Perhaps the first to make a major contribution was a general in Napoleon Bonaparte's army, Count Pierre Dejean. Dejean laid the foundation for the systematic organization of tiger beetles, and many of his groupings are still valid today. Building on Dejean's foundation was the work of the German-born entomologist Walther Hermann Richard Horn. He accepted most major groupings proposed by Dejean and others before him, and he split tiger beetles into two main groups. This division into two lineages was his attempt to reflect their evolutionary relationships to each other and to other groups of insects. Following Horn was the work of Emile Rivalier, who studied the shape and structure of male tiger beetle genitalia. He published several papers from the 1950s into the 1970s, each focused on a major biogeographic region. Based on similarities he observed, Rivalier was able to elevate over fifty subgenera. Many of his subgeneric classifications have since been elevated to full generic status. Recently, DNA has been used to reevaluate the relationships of North American tiger beetles, and many of Rivalier's groupings are supported. Those not supported have been placed in new groupings using a combination of DNA evidence, morphology, biogeography, and ecology. We have chosen to follow this new classification as presented by Duran and Gough (2019) for this publication.

Currently there are more than 2,700 species of tiger beetles described worldwide. In North America, there are 116 species with an additional 153 recognized subspecies (species and subspecies are collectively known as taxa). The southeastern United States is home to 52 taxa representing 8 genera. The genus *Tetracha* (Metallic Tiger Beetles) is represented by three taxa in the region. Formerly considered a subgenus of *Megacephala*, these are the largest tiger beetles in the Southeast. *Tetracha* are flightless and primarily nocturnal, and they can often be found in disturbed habitats such as lawns and agricultural fields. Our largest genus, with 15 taxa, is *Cicindela* (Temperate Tiger Beetles). Members of this genus can be found in a variety of habitats including river bars, woodland trails, and open sandy upland areas. The remaining tiger beetles in the Southeast have over time been removed from the genus *Cicindela* and placed into a number of other genera. The genus *Cicindelidia* (American Tiger Beetles) is represented in the Southeast by 10 taxa. These small to medium-sized tiger beetles occupy a wide variety of habitats both near and far from water, and the genus includes one of our most widespread and ubiquitous as well as several of our rarer species. The genera *Apterodela* (Leaf Litter Tiger Beetles) and *Parvindela* (American Diminutive Tiger Beetles), each represented by one taxon in the region, include one of our larger and one of our smaller tiger beetles, respectively. Both are flightless (or only rarely fly) and typically occupy less open habitats than other tiger beetles found in the area. Ten taxa of southeastern tiger beetles belong to *Ellipsoptera* (Ellipsed-winged Tiger Beetles). Members of this genus occupy a variety of habitats including coastal and bay shores, river banks and sandbars, and sandy upland areas with open canopies and sparse vegetation. Four taxa are included in *Eunota* (Saline Tiger Beetles); these tiger beetles may be found along brackish canals as well as in salt marshes and salt pannes. A single variable species comprised of three taxa is included in the southeastern *Habroscelimorpha* (Habro Tiger Beetles); this species is restricted to ocean and bayside coastal shores. *Brasiella* (Little Tiger Beetles) and *Microthylax* (Coral Beach Tiger Beetles) are each represented by one taxon in the southeastern United States. These tiger beetles are not native to our area and have only been recorded in the Florida Keys, probably having been blown in from Cuba during storms. Neither is currently known to be extant in the Southeast.

As researchers and other interested workers gain a better understanding of tiger beetles through new scientific techniques, their placement within a genus, or even at a species or subspecies level, may be changed to better represent their taxonomic relationships. In some cases a new genus or subgenus name will need to be given to a taxon to reflect this change in placement; the older, outdated names are referred to as synonyms. For example, the tiger beetle currently known as *Tetracha carolina* was originally described by Linnaeus as *Cicindela carolina*. It was later transferred to the genus *Megacephala* (and was called *Megacephala carolina*); it was subsequently placed in the subgenus *Tetracha*. At that time its complete scientific name would have been *Megacephala* (*Tetracha*) *carolina*, and the name *Cicindela* would be a synonym for this species. Then, in 2007, Roger Naviaux revised the genus *Megacephala*, elevating the subgenus *Tetracha*, including the North American species, to full generic status,

Size comparison of southeastern US tiger beetles: (*left to right*) Virginia Metallic Tiger Beetle (*Tetracha virginica*), Carolina Metallic Tiger Beetle (*T. carolina carolina*), Six-spotted Tiger Beetle (*Cicindela sexguttata*), Bronzed Tiger Beetle (*Cicindela repanda repanda*), S-banded Tiger Beetle (*Cicindelidia trifasciata ascendens*), Punctured Tiger Beetle (*Cicindelidia punctulata punctulata*), Eastern Beach Tiger Beetle (*Habroscelimorpha dorsalis saulcyi*), Eastern Pinebarrens Tiger Beetle (*Cicindelidia abdominalis*), Ant-like Tiger Beetle (*Parvindela cursitans*).

which resulted in a change of the scientific name to *Tetracha carolina*. Over time, then, *Tetracha carolina* has been referred to as *Cicindela carolina, Megacephala carolina*, or *Megacephala* (*Tetracha*) *carolina*, depending on the age of the literature. Often included with the name is the authority—the person to first describe the species. Parentheses around the authority indicate that the original combination of species and subspecies has changed from what it was when first described. We can again use *Tetracha carolina* as an example. As first described, it would appear as *Cicindela carolina* Linnaeus, and all subsequent combinations would include Linnaeus in parentheses, i.e. *Tetracha carolina* (Linnaeus). Even this basic familiarity with synonymy and authorship should assist the uninitiated when searching older literature or viewing specimens in museums or collections that either haven't adopted the most current taxonomy or haven't updated their labeling system to reflect the most recent taxonomic changes.

Finding and Observing

Just like with many other groups of insects, the key to finding tiger beetles is habitat—if you aren't in the right habitat, you are not going to find very many. After reviewing the habitat section and/or the species accounts, this section should help you find and observe tiger beetles once you are in the right spot.

After being in the right habitat, the next most important factor is probably weather conditions. Below a certain temperature, tiger beetles will spend most of their time either underground in burrows or

Clockwise from top left:

Oblique-lined Tiger Beetle (*Cicindela tranquebarica tranquebarica*) ground hugging for warmth. Towns County, Georgia.

Splendid Tiger Beetle (*Cicindela splendida*) stilting for thermoregulation. Towns County, Georgia.

Punctured Tiger Beetle (*Cicindelidia punctulata punctulata*) using vegetation to avoid surface layer heat. Appling County, Georgia.

hidden in dense vegetation, in leaf litter, under logs, etc. This threshold temperature varies by species, with cold-adapted species (those with a more northern affinity) appearing earlier in the year and at much cooler temperatures than other species. Species that spend the winter as adults may come out on unusually warm days as early as February in the South; if it's sunny, a daytime temperature of 50°F may be warm enough, although 60°F is better. In general, tiger beetles will come out sooner on sunny days; it may take a 10-degree temperature increase on a cloudy day to get the same level of activity. Conversely, at the end of the season when tiger beetle populations are dwindling, hours of activity may be limited, and when the sun drops low in the afternoon, they may vanish. For some species, daytime temperatures in the summer may get too warm, and activity will cease in midday as the beetles seek shelter from the sun, usually by digging into either temporary or semipermanent burrows.

Tiger beetles have a number of thermoregulatory behaviors they can use to ameliorate less than ideal conditions. To warm themselves when temperatures are too cool, behaviors include basking in the sun or lowering their bodies to the ground (if the substrate is warmer than the air). This behavior is called ground hugging.

When air temperatures are too warm, tiger beetles may assume a posture on fully extended legs to get their body as high as possible; this posture is called stilting. This behavior allows them to avoid the surface layer of air next to the ground, which is often the hottest.

Less commonly, beetles may climb just above the ground on low vegetation for the same reason.

Finding shade, usually in vegetation or beneath fallen logs, is another behavior used to escape high temperatures. In species that inhabit areas near streams or ponds, individuals may move toward the wetter, cooler sand along the water's edge as temperatures increase. Longer continuous periods of high temperatures usually cause the beetles to retreat to their existing burrows or to dig temporary burrows, which they can do remarkably rapidly, especially in sand. During midsummer, many of the species that inhabit dry sand areas will be underground by 9:00 or 10:00 a.m. and may only reappear for a short time in late afternoon.

Once you have found a suitable habitat, in the right season, and at the right temperature, you are ready to find some beetles! Most tiger beetles are found in relatively open habitats, and they typically run around in short bursts of movement followed by a pause. You can either stand in one place and watch for movement on the ground, or you can walk slowly along and watch for individuals to run ahead of you or make short flights to escape what they perceive as possible danger. Because many species will land and just stand still for a minute or two, if you are able to watch the flight and note where they land, and if you approach slowly without making any sudden movements, you may be able to walk right up to them. Crouching to create a lower profile helps as you try to get closer, and approaching from the rear is also easier because a tiger beetle's vision is weakest in that direction. Some species will fly into vegetation when disturbed, so those will take a little more patience as you wait for them to reappear where you can see them. In general, it is much easier to look for movement than it is to specifically search for a tiger beetle or beetle shape, so while searching, you will likely see a lot of other moving creatures, including ants, spiders, and especially flies. In time you will learn what a running tiger beetle looks like (or, more often, what a tiger beetle's escape flight looks like!), and it will get easier to ignore other movement and focus on tiger beetles. Close-focusing binoculars will make observing tiger beetles much more rewarding; they can be the only way to really see those species that are extremely wary and hard to approach no matter how slow and stealthy you are.

Try to focus on the specific part of the habitat where a tiger beetle might be found. For example, while searching for river species, look for stretches of sandier open ground, sandbars, or areas of sand that have been deposited up on the bank. While searching along forest paths, unless it is very warm, you will find more individuals in sunny open spots. Some, like the Six-spotted Tiger Beetle (*Cicindela sexguttata*), will fly from one sunny spot to another, avoiding shaded areas (unless they are trying to escape from you, then they'll fly into the darkly shaded forest, never to be seen again). As discussed earlier, if it gets too warm, many species will be found almost exclusively in the shade and not in the sun.

Metallic Tiger Beetles (*Tetracha* spp.) and other nocturnal species are harder to find because they rest beneath cover during the day. To find them search just about any open or semi-open habitat—the edges of agricultural fields are great places—looking under the boards or small logs where these beetles hide during the day. Always replace the

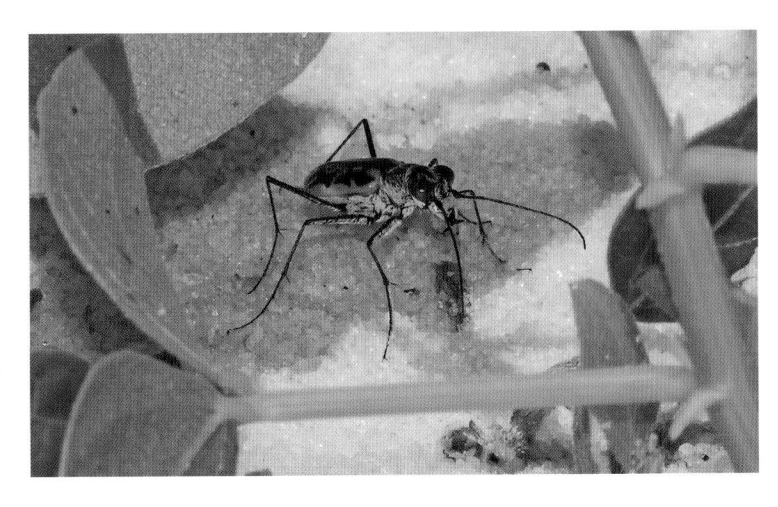

Coastal Tiger Beetle (*Ellipsoptera hamata lacerata*) in shade to avoid the heat. Franklin County, Florida.

log or board after turning it over. Other strategies for finding nocturnal species include going out at night with a flashlight and actively hunting for them, or setting up UV/mercury vapor lights at night to attract them.

Finding tiger beetle larvae can also be fascinating, and it's even more critical that you are exactly in the right habitat. This may sound obvious, but it's easier to find the burrows than it is to look for the larvae. As you walk along through the habitat, the vibrations from your footfalls will send the larvae down into their burrows for safety; the key is to recognize which of the myriad of small open holes in the dirt or sand might be a tiger beetle burrow. One hallmark of tiger beetle larval burrows that is different from burrows of most other burrowing invertebrates is that the pellets from excavating are always cast in the same general direction; instead of a ring of pellets around the hole, there is a pile just on one side. This is an excellent way to quickly hone in on which burrows are those of tiger beetles.

Many species of spiders dig burrows that are about the same size as tiger beetle burrows, but they will either have no pellets around them or have pellets more or less evenly distributed around the

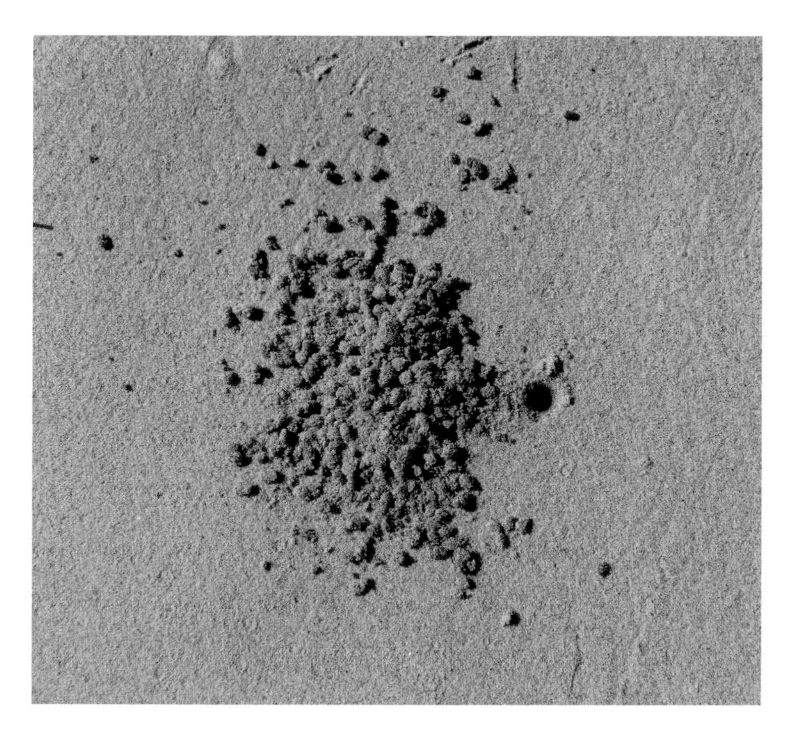

Larval tiger beetle pellet throw pattern. Early County, Georgia.

entrance. Once you have found a burrow that you think may belong to a tiger beetle larva, find a comfortable place to sit or stand that does not shade the burrow entrance and wait. If the larva was actively feeding or excavating before sensing your approach, they will usually resume the activity in 5 or 10 minutes; it's important that you keep very still and don't shift around, because every time you move you create more vibrations that will alert the larva to your presence. Although they often come up only momentarily the first couple of times, if you wait, the larva will eventually resurface and resume feeding by staying at the top of the burrow; if you remain still, you can then observe them as long as you want.

Photographing

The good news is that there are only two keys to photographing tiger beetles. The first—and most important—is to get close enough to use whatever equipment you have, and the second is to know how to use that equipment and be set up before you get close to your subject, because you may not get many attempts before the beetle runs or flies off. That's it! The bad news is that the first of these isn't always easy, but it is usually possible if you are patient and persistent. Having binoculars with you will also help; you can look at each beetle before you start your stalk to see what it is you are stalking. For example, a river sandbar may have hundreds of Bronzed Tiger Beetles (*Cicindela repanda*) but only a handful of Twelve-spotted (*C. duodecimguttata*) or Hairy-necked (*C. hirticollis*) Tiger Beetles; if you are trying for one of the latter, you'll probably want to know you've got the right species before you start to sneak up on one.

The same tips in the Finding and Observing section work for photography because your goal is the same: to get as close as possible without flushing the beetle or having it run off. Standard tiger beetle stalking tips include locating your subject at a distance and trying to move slowly into position to shoot. One crucial difference in photography is that you need to be aware of the sun at all times—there's nothing worse than completing a long, successful stalk, getting in position, and then having to shoot into the sun. While looking for subjects, walk with the sun at your back. When you reach the end of the habitat, walk back to the beginning without trying to find anything to photograph, and then start again with the sun behind you. Morning and late afternoon generally provide better light for shooting tiger beetles, since a strong overhead sun washes out many color details. Days with a high, thin overcast are generally going to be the best days for photography, as long as the temperature is high enough to allow lots of beetle activity. A dark, cloudy day rarely provides many subjects to photograph.

Since many tiger beetles are very shiny and highly reflective, it's difficult to get shots that aren't at least somewhat "blown out" by highly reflective spots somewhere on the beetle. Shooting on lightly overcast days helps, but many photographers try to create the same effect of softening and evening the light by shading the beetle with a portable diffuser. Diffusers are available in many sizes, but those larger than about 12–15 inches are pretty unwieldy to use outdoors, especially if there is any breeze at all. The diffuser cuts down on the harshness of the light, allowing you to better capture the fine details of the beetle. If you have a friend or partner who is willing to go out with you, they can hold the diffuser, and you only have to worry about getting in position and setting up the camera for the shots. Another successful technique involves shading the beetle with your body, a hat, or some other article that completely blocks the sun. The main problem with all of these light-blocking or diffusing attempts is that many tiger beetles will not sit still while being shaded. You will have the best results if you move the shade (diffuser, hat, body, etc.) slowly over the beetle, like a passing cloud. Even then many will run or fly off, so you'll have to go find another subject and try again. Some species are so wary you may have to do this repeatedly until you find one that is willing to sit still and be photographed. Once you have found a beetle willing to remain stationary in the shade, you will have to dramatically slow your shutter speed in order to still get enough light for the image; you may need to use speeds as slow as 1/10 second for fully shaded tigers. Although it's not easy to get these hyperactive insects to sit still long enough to be photographed using these techniques, when it all comes together, the resulting images are wonderful and well worth the effort.

Depth of field isn't as big a problem with tiger beetles as it is with some insects, and a setting anywhere between about f/8 and f/22, depending on light conditions and the personal preference of the photographer, should provide suitable depth of field for a pleasing image.

As far as equipment goes, there are many combinations of cameras and lenses that can result in good tiger beetle images. However, there are about three major setups that most field insect photographers use.

The first, and easiest, is a good point-and-shoot camera with a zoom lens, which allows you to shoot decent images without having to get too close to the beetle. A few of the shots in this book were taken with a handheld Lumix FZ-1000 camera, usually with the tiger beetle being shaded by the photographer or a helper. A good point-and-shoot camera like this one can produce really nice sharp shots, even when hand held.

The other two choices both involve Digital Single-Lens Reflex (DSLR) camera bodies and interchangeable lenses. Using very long focal length lenses in the 300–500 mm range, you can shoot tiger beetles without having to be right on top of them, but it's a trade-off. You don't have to get as close, but you can't control the light without a helper, either. Many photographers use handheld 300- or 400-mm zoom lenses and a fill flash; this helps to make up for any camera shake by speeding up the shutter speed. However, we do not think that fill flash works consistently well on highly reflective subjects like tiger beetles, as the resultant images often have hotspots and other blown out areas with too much light. For this reason we generally do not use fill flash on tiger beetles and don't recommend it, unless the conditions or subject are so dark that you cannot get a usable image any other way. One note here about fill flash: if you are forced to shoot tiger beetles in full sun, just a small amount of diffused fill flash (dial down the power on your flash and use some sort of diffuser) can bring back a few of the details you lose in the full sun.

The following option is what we used to take almost all of the photos in this book, and the technique that we feel provides the best overall tiger beetle images. It's also the most demanding, because one of the key items is a tripod. Sneak up on your subjects and shade them; then, with DSLRs and true macro lenses (100–200 mm) mounted on tripods, use long exposures (very slow shutter speeds) to allow for sufficient light in each image. You will also need a remote shutter release for this technique; at these slow speeds, even on a tripod you will move the camera just by pressing the shutter button. This may not be the technique to start your career shooting tiger beetles, but if you have been shooting them for a while, we recommend it for the most evenly lit images with good colors and no blown highlights. And in case you are wondering how this works, you load your camera on the tripod, sling the tripod on your shoulder, and trudge off hoping to find willing subjects. Obviously, the lightest tripod you can obtain is a help here! One other option is that some companies offer low boy tripods with legs that only extend to a maximum of 12–18 inches; since there isn't much leg, there isn't as much weight, either. Another feature to look for in a tripod is the ability to get right on the ground for ground-level shots. Most tripods accomplish this in one of two ways: no center section on the tripod and legs that extend out at 90 degrees, or some gizmo above the legs that lets the center section of the tripod flip out to the side so you can lower it close to the ground. For identification purposes, most of our shots for this book were taken slightly above ground level to show both dorsal and lateral features; low to the ground shots can be very interesting, though, so try to get a tripod that can do them as well.

Conservation

Historically, conservation efforts for insects have received little attention compared to the charismatic megafauna such as birds and mammals. In 1994 only 29 insects in the United States were federally listed, compared to 335 mammals, 112 reptiles, and 82 mollusks. However, in recent years that number has more than doubled, with 76 species currently listed as either threatened or endangered. This trend reflects the importance government agencies and nongovernmental organizations are now placing on this once overlooked group.

According to data from the Xerces Society (2016), tiger beetles are the second-most endangered group of insects, with 19 percent of taxa considered vulnerable, imperiled, or extinct. Habitat destruction and degradation are the primary factors threatening tiger beetle populations today. This is particularly true for habitat specialists and those species with highly localized populations. For upland species, conversion of land for agricultural and municipal purposes directly destroys and fragments habitat, and exposes affected tiger beetle populations to associated threats such as pesticides and light pollution. Many species are attracted to light and congregate at artificial sources, making them easy targets for opportunistic predators. In coastal areas, dredging and filling of estuarine habitats and shoreline modifications (bulkheads, seawalls, riprap, etc.) have been implicated in the decline of a number of populations. Natural wildfires were once a common occurrence in the Southeast and kept forests relatively open and free of woody understory, especially in pine forests. However, fire suppression has allowed many of these areas to become overgrown and unsuitable for tiger beetles.

Riparian species are at risk as well. When dams are constructed on rivers, habitats upstream are lost to flooding, and altered flow regimes downstream reduce the scouring effect periodic floods have on minimizing the vegetative encroachment of river bars. Increased disturbance due to foot traffic and vehicular use in any of these areas can negatively impact tiger beetle populations in a number of ways. At worst, the increased chance of direct contact can potentially crush adults and larvae. Indirectly, soil compaction may render previous areas of habitation unsuitable, and the increased disturbance interferes with natural behaviors such as feeding, mating, and oviposition (Knisley 2011).

Anthropogenic factors are not the only threats to tiger beetle populations. Random catastrophic natural events such as severe storms and prolonged flooding are a real threat for species with small, highly disjunct populations. The more disjunct a population is, the less likely a neighboring population will be able to recolonize if any suitable habitat remains intact. Coastal species are particularly prone to catastrophic natural events (Knisley 2011).

There are currently five US taxa of tiger beetles listed as threatened or endangered, one of which occurs in the region. The Miami Tiger Beetle (*Cicindelidia floridana*) is a Florida endemic that occurs only in Miami-Dade County. The species was long thought to be extinct, having not been seen since its discovery in 1934. It was first described as a variety of the Eastern Pinebarrens Tiger Beetle (*C. abdominalis*) and was later considered a subspecies of the Scabrous Tiger Beetle (*C. scabrosa*), before being elevated to a full species after its rediscovery in 2007. This tiger beetle occupies a very small area of pine rockland habitat in the Miami metropolitan area that is under immediate threat from development and encroachment of nonnative invasive plant species. The US Fish and Wildlife Service listed the Miami Tiger Beetle as endangered on November 4, 2016.

A number of other southeastern tiger beetle species have seen a decline in recent years. The Highlands Tiger Beetle (*Cicindelidia highlandensis*) was under consideration for federal protection until a recent ruling deemed that listing was not currently warranted. It is endemic to central Florida and is known only from Polk and Highlands counties along the Lake Wales Ridge. The area is typically scrub habitat with deep sandy soils and is known for its high degree of endemism. Unfortunately, citrus grows well in the area and much of the natural habitat has been converted to groves. The Cobblestone Tiger Beetle (*C.*

marginipennis) was historically known in the region from Mississippi and Alabama, but it is now believed to be extirpated from Mississippi. In Alabama it is apparently restricted to a handful of sites along three large rivers in the central part of the state; there are no recent records from one of these rivers, and the species may no longer occur there. Another riparian species under threat is the Appalachian Tiger Beetle (*Cicindela ancocisconensis*), which in the Southeast is apparently now confined to a couple of watersheds in North Carolina and Tennessee. Both of these riparian species are especially vulnerable to stream channel impoundments and other alterations, as well as water quality degradation. The Ghost Tiger Beetle (*Ellipsoptera lepida*) has a broad distribution ranging from New York to Nevada, where it primarily occupies dry, well-drained sandy soils of inland dunes, sandy washes, and sand ridges. In the Southeast it primarily occupies the upper reaches of large river sandbars. These sandbars are susceptible to encroachment by vegetation if periodic flooding is mitigated. Another widely distributed species occurring in the Southeast is the Hairy-necked Tiger Beetle (*Cicindela hirticollis*). Here it occupies coastal beaches as well as river banks. While coastal populations appear to be stable, those inland are becoming increasingly difficult to encounter, possibly due to various river modifications such as damming, channelization, and shoreline development. The Northern Barrens Tiger Beetle (*C. patruela*), an inhabitant of dry sandy soils in mixed oak and pine forests, and the Splendid Tiger Beetle (*C. splendida*), an inhabitant of open red clay soils, have both experienced population declines in the Southeast, to the extent that the former is almost extirpated from the region. Fire suppression, as well as habitat destruction and degradation, appear to be the primary causes for their decline.

While some species of tiger beetles are the subject of conservation concern, others may be used as conservation tools. Aquatic invertebrates have long been used as bioindicators for aquatic systems. It is only recently that scientists have looked to their terrestrial counterparts to fulfill a similar role. These bioindicators may be used to represent other organisms in biodiversity estimates, and they also give us an idea of the overall health of an ecosystem—the proverbial canary in a coal mine. An ideal bioindicator must meet several criteria, including but not limited to: classification is well-known and stable; species is easily identified; species is easily observed in the field; and life history and biology are well-known. Tiger beetles fit these criteria quite well and have even been used to predict bird and butterfly diversities (Pearson and Vogler 2001).

Collecting

As discussed elsewhere, most adult tiger beetles in the southeastern United States can be identified with a good pair of close-focusing binoculars or a quality photograph. However, there are a few instances when an individual will need to be observed more closely in order to make a positive identification. In these cases, and for those folks who might want to make a reference collection of specimens, it will be necessary to capture a wary insect.

The techniques for capturing tiger beetles, and for preparing and storing a reference collection, have been discussed in detail in many publications (such as Pearson et al. 2015), so we'll just hit the highlights here. First and foremost, it is very important to be sure that collecting is permitted in your location. For much public land in the Southeast, a scientific collecting permit is required to keep specimens, although this varies by state (and sometimes within a state, depending on which state or federal agency manages the property). Check with the appropriate agency to make sure that it's permissible to collect on the property you're visiting. If the land you're on is private, then it is certainly advisable to get verbal permission from the landowner before collecting any tiger beetles.

Usually the most difficult part of the collecting process is being able to approach within net range of the beetle! The same techniques for getting close to tiger beetles discussed in the Photographing section will work just as well here. Once you've successfully approached an individual, the most often utilized equipment for collecting adult tiger beetles is the standard aerial insect net (also known as a butterfly net). There are two basic techniques employed when using a net to collect tiger beetles: (1) pancaking—slapping the net down on the ground, trapping the beetle underneath, and capturing it before it can escape from one of the inevitable gaps where the rim touches the ground; or (2) flushing the beetle into flight and then sweeping from behind to capture it in the net bag. On windy days, this second technique works best if you approach the beetle from downwind; when it launches into flight, the wind will help push it toward your net instead of away from it. The habitat conditions—and skill of the collector—often dictate which of these methods will be more successful. For the few flightless species in the region, it is often easiest to trap them on the substrate with your hands rather than using a net. Some species, especially those that are most active very late in the day and into the night, can be attracted to black lights at night and can be most easily collected this way. Other collecting methods, such as pitfall traps, are more specialized in nature and will not be discussed here.

Collecting larval tiger beetles is a very specialized skill, requiring great care and patience. It involves either carefully digging them out of their burrows, fishing them out with a stem of grass or similar object, or possibly a combination of the two techniques. For those interested in more information on collecting or curating tiger beetle larvae, we recommend the excellent discussion in Pearson et al. (2015).

If the creation of a reference collection is the goal, then the most important thing to remember is to always keep a complete set of data with the specimen at all times; specimens without data are practically useless! These data should include, at a minimum, the date of collection, location (state, county, specific locality, latitude/longitude), and collector. Once the specimen is identified, that information may also be added. In terms of actual curation, a summary of our method would be to preserve the adult tiger beetle in ethyl alcohol (where it can be stored until you are ready to proceed); transfer the beetle to an organic solvent (such as acetone) for a week or two to remove some of the fats that can otherwise obscure the color patterns; relax the specimen, either by placing it in a closed, humid environment for a short period or by soaking in a fluid designed for this purpose; pinning the beetle; and storing (with data) the pinned specimen in a collection drawer, in low humidity and away from insect pests. This procedure results in a properly curated tiger beetle that can be donated to a research collection at a later date, if appropriate.

How to Use the Species Accounts

In this book, we describe a total of 52 identifiable taxa and forms. We have included the 42 species of tiger beetle that have been recorded from the states that make up the region covered: Alabama, Florida, Georgia, Mississippi, North Carolina, South Carolina, and Tennessee. For species with recognizable subspecies, we have also included every subspecies known to occur in the region. Two species previously found in Florida, but not currently known to exist there, are included in the appendix.

Each account provides the common and scientific names for the species and information regarding size, identification, similar species, Southeast status, and habitat, plus notes, photographs, a range map, and a flight season bar graph. The common (Pearson et al. 2015) and scientific names are followed by the name of the author who formally described the species. Size ranges are based on specimens measured directly by the authors, published in reference articles and books, or both. We have also attempted to give subspecies size differences when we could obtain that data. The identification section describes in detail the size, colors, and maculations of the species and all regional subspecies. Every species follows the same format to make it easier to compare one species with another. We emphasize features that are visible in the field or in the hand. The similar species section lists and describes the species we feel are most similar to the species in the account. Identifying features and behaviors for similar species are listed here, along with the best ways to distinguish each species.

The Southeast status is simply a short description and general guide of how common (or uncommon) and widely distributed the species is. In some cases the relative rarity of the tiger beetle is largely determined by the rarity of the habitat, such as the Gulfshore Tiger Beetle (*Eunota pamphila*) and salt pannes. In all cases you must find the correct habitat or it doesn't matter how common a species may be, you won't find it. The habitat section describes the type of habitat where the species is found. Keep in mind it is difficult to succinctly describe all the microhabitats within a region or where each species of tiger beetle may be found, but we have tried to keep this section both short and useful.

The range maps are based on every county-level record we could find in the region. We used literature searches, museum searches, online databases, and (in a few cases) private collection searches, plus our own fieldwork. We have undoubtedly missed some valid records, but every record we could find and verify is included. In a few cases, published records that are very questionable and that we were unable to verify have been omitted from the maps. Users of this book are encouraged to send us or make publicly available any additional records from this region. Note that these are county-level maps, so one verified record within a county, no matter how large or small the county is, results in the entire county being marked. We gave greater scrutiny to records that were outside the known ranges for a species, so every map with an outlier record mapped has had that record verified.

The season bar graph shows the approximate flight season for each species. The darker central portion of each graph indicates the peak of the flight season. We have averaged the emergence dates for the entire region in the bar graph, so an observer in Florida may see a species slightly before the graph indicates, and an observer in the Blue Ridge of Georgia, Tennessee, or North Carolina may only find a species later than the graph indicates.

We have used the note section to include any other useful information about behavior, including general wariness for each species, and any other useful or interesting information about the taxonomy or natural history of the species. Observers will soon learn that some species, like the One-spotted Tiger Beetle (*Apterodela unipunctata*), are very easy to approach, while others are incredibly hard to get close to—the Cobblestone Tiger Beetle (*Cicindelidia marginipennis*) surely fits in the latter category! We have tried to give users a sense of how easy it will be to approach each species, because this knowledge will change how you look for them in the field. If you are only scanning the ground a few feet ahead of you as you walk along, very wary species like the Cobblestone Tiger Beetle, which will be flushing ahead of you from farther away, might never be seen!

Species Accounts

Carolina Metallic Tiger Beetle

Tetracha carolina carolina (Linnaeus)

Size 14–18 mm.

Southeast Status Fairly common throughout region except absent from south Florida.

Identification A large, robust tiger beetle with metallic multi-colored granulate elytra. The head is green with variable amount of purplish, coppery, or orange highlights along the center of the dorsum. The antennae and labrum are a pale-yellowish color. The pronotum is metallic green with purple or copper highlights. The elytra are obviously granulate, with a wide metallic-green margin. The center of the dorsal surface is a shiny purple or coppery on the anterior half, fading to a dark almost black posteriorly. The only maculation is the apical lunule, expanded into a large U-shaped pale-yellow or cream-colored marking at the rear of the elytra. The U-shaped mark on the anterior ends is round, expanded, and slightly convergent toward midline. The apical lunule is much larger than typical tiger beetle apical lunule. The legs are pale yellowish, and the abdomen is brownish yellow.

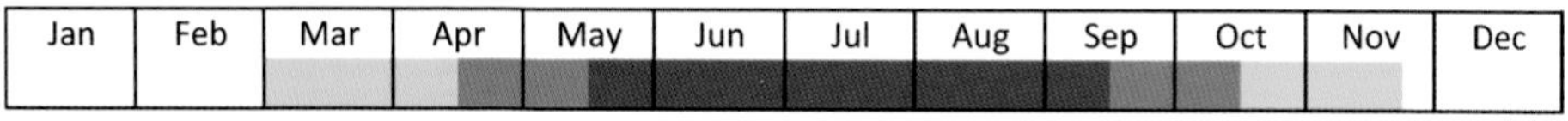

Jan	Feb	Mar	Apr	May	Jun	Jul	Aug	Sep	Oct	Nov	Dec

Above: Flight season of the Carolina Metallic Tiger Beetle (*Tetracha carolina carolina*).

Left to right:

Carolina Metallic Tiger Beetle (*Tetracha carolina carolina*). Coosa County, Alabama.

Carolina Metallic Tiger Beetle (*Tetracha carolina carolina*). Murray County, Georgia.

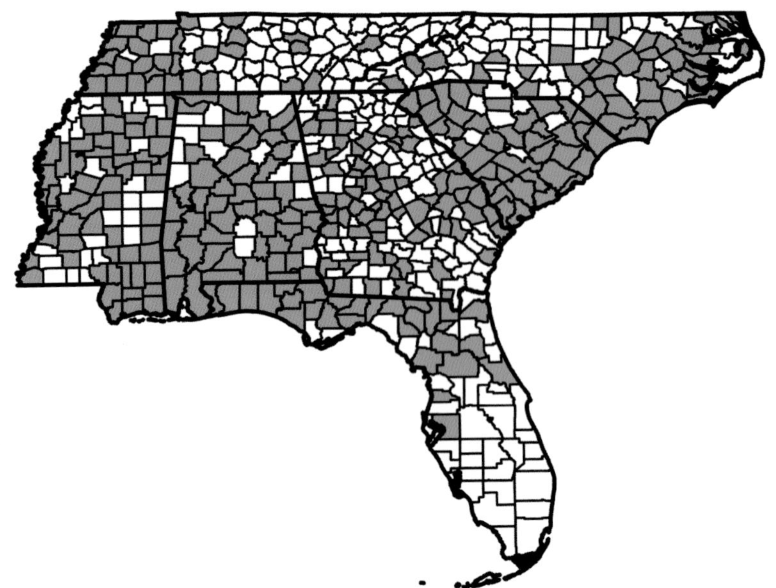

Range of the Carolina Metallic Tiger Beetle (*Tetracha carolina carolina*) in the southeastern United States.

Similar Species The Carolina Metallic Tiger Beetle can only be confused with other members of this genus. The Virginia Metallic Tiger Beetle (*T. virginica*) is on average slightly larger, shiny green and black with no purple or copper colors, and without the U-shaped elytral mark. The slightly smaller Florida Metallic Tiger Beetle (*T. floridana*) is very similar but without the purple or copper color; all its surfaces are green or blackish. It has a similar U-shaped mark that is slightly darker in color but does not converge toward midline and has minimal rounded tip expansion. The Florida Metallic Tiger Beetle is restricted to south Florida.

Habitat Utilizes many different habitats, but especially common along rivers, streams, marshes (fresh or salt), and other water-edge habitats. Also found in moist grassy areas such as fields and pastures as well as around agricultural fields and other disturbed habitats.

Note When found foraging, this nocturnal species is usually running and difficult to capture or photograph. It almost never flies, although it can. Can be found under boards or other debris during the day, sometimes several individuals together. Often attracted to lights, sometimes in numbers. Exact status somewhat obscured by nocturnal habits. The limits of range are not completely known in south Florida due to similarity to, and confusion with, Florida Metallic Tiger Beetle.

Florida Metallic Tiger Beetle

Tetracha floridana (Leng and Mutchler)

Size 12–16 mm.

Southeast Status Fairly common in south Florida, including the Florida Keys, with a few records as far north as Dixie County.

Identification A large, robust metallic-green and black beetle. The head is a shiny green. The antennae, labrum, and legs are a pale-yellow color. The pronotum is metallic green with a small amount of black at the anterior edge. The elytra are granulate, with metallic-green margins. The dorsal surface is centrally dark, almost black. A large, pale apical lunule is present and much larger than typical tiger beetle apical lunule; the anterior ends are slightly expanded and rounded, and typically do not converge toward the midline.

Jan	Feb	Mar	Apr	May	Jun	Jul	Aug	Sep	Oct	Nov	Dec

Above: Flight season of the Florida Metallic Tiger Beetle (*Tetracha floridana*).

Clockwise from top:

Florida Metallic Tiger Beetle (*Tetracha floridana*). Pinellas County, Florida.

Florida Metallic Tiger Beetle (*Tetracha floridana*). Pinellas County, Florida.

Florida Metallic Tiger Beetle (*Tetracha floridana*). Lee County, Florida.

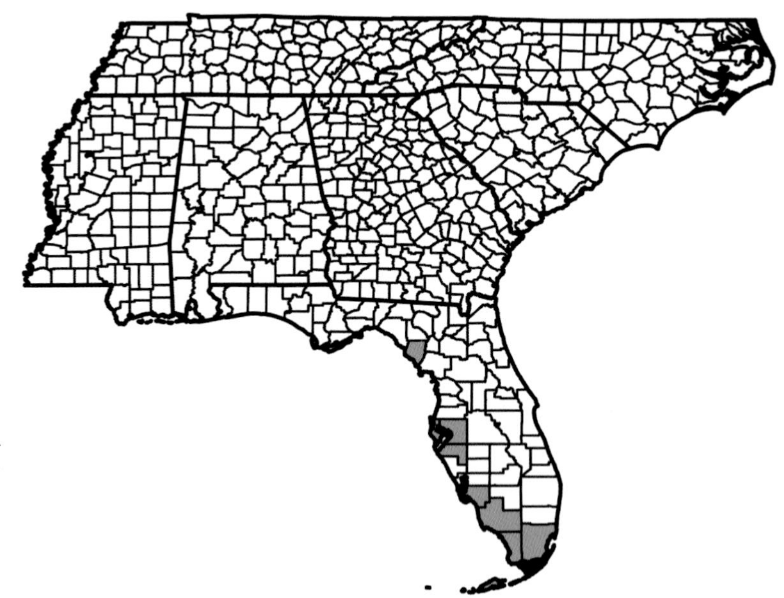

Range of the Florida Metallic Tiger Beetle (*Tetracha floridana*) in the southeastern United States.

Similar species The Florida Metallic Tiger Beetle can only be confused with the very similar Carolina Metallic Tiger Beetle (*T. carolina*), which is slightly larger, has purple or copper coloration on most dorsal surfaces, and has an apical lunule that abruptly expands at the anterior end and converges toward the midline. Their range overlaps only at the southern end of the Carolina Metallic Tiger Beetle range.

Habitat Most easily found on alkaline mudflats of the Florida Keys and extreme south Florida. Also a few records from saline marshes and adjacent wet, grassy, or mudflat areas along the Gulf Coast. Often associated with succulent vegetation such as glasswort (*Salicornia* sp.).

Season Summer and fall, with some reports suggesting it is active throughout the year (Pearson et al. 2006).

Note Nocturnal and a fast runner. When caught in a flashlight beam, instead of running away as other *Tetracha* species do, often runs erratically in tight circles. Exact range limits in peninsular Florida obscured by similarity to, and confusion with, Carolina Metallic Tiger Beetle.

Virginia Metallic Tiger Beetle

Tetracha virginica (Linnaeus)

Size 16–21 mm.

Southeast Status Fairly common throughout the region.

Identification A large, robust metallic-green and black beetle. The head is green. The antennae, labrum, and legs are a pale-yellow color. The pronotum is metallic green with a black center. The elytra are granulate and centrally black with metallic-green margins. It has no pale markings.

Similar Species Distinctive; lacks the large, pale apical lunules present in other members of this genus.

Jan	Feb	Mar	Apr	May	Jun	Jul	Aug	Sep	Oct	Nov	Dec

Above: Flight season of the Virginia Metallic Tiger Beetle (*Tetracha virginica*).

Clockwise from top:

Virginia Metallic Tiger Beetle (*Tetracha virginica*). Bibb County, Alabama.

Virginia Metallic Tiger Beetle (*Tetracha virginica*) daytime refugium. Tattnall County, Georgia.

Virginia Metallic Tiger Beetle (*Tetracha virginica*). Towns County, Georgia.

Range of the Virginia Metallic Tiger Beetle (*Tetracha virginica*) in the southeastern United States.

Habitat Almost any open or partly open habitat, including river and lake edges, grassy areas, and developed areas.

Note Because this flightless species is primarily nocturnal and a rapid runner, it is difficult to capture or photograph. Can be found under boards or other debris during the day. Not as common as Carolina Metallic Tiger Beetle (*T. carolina*) but can be locally abundant.

One-spotted Tiger Beetle

Apterodela unipunctata (Fabricius)

Size 14–18 mm.

Southeast Status Fairly common throughout most of the region; absent from Florida and uncommon to rare in the Atlantic Coastal Plain.

Identification A large, dull-brown tiger beetle with minimal markings. The head, pronotum, and elytra are a mottled dark brown. The labrum is large and white. Markings are limited to a small marginal white spot at the middle band of each elytron; spot is often absent or muted. The legs are brown or metallic green, and the abdomen is metallic green.

Jan	Feb	Mar	Apr	May	Jun	Jul	Aug	Sep	Oct	Nov	Dec

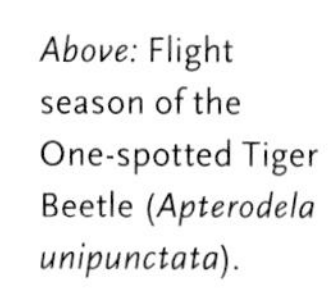

Above: Flight season of the One-spotted Tiger Beetle (*Apterodela unipunctata*).

Clockwise from top left:

One-spotted Tiger Beetle (*Apterodela unipunctata*). Lumpkin County, Georgia.

One-spotted Tiger Beetle (*Apterodela unipunctata*). Dawson County, Georgia.

One-spotted Tiger Beetle (*Apterodela unipunctata*). Towns County, Georgia.

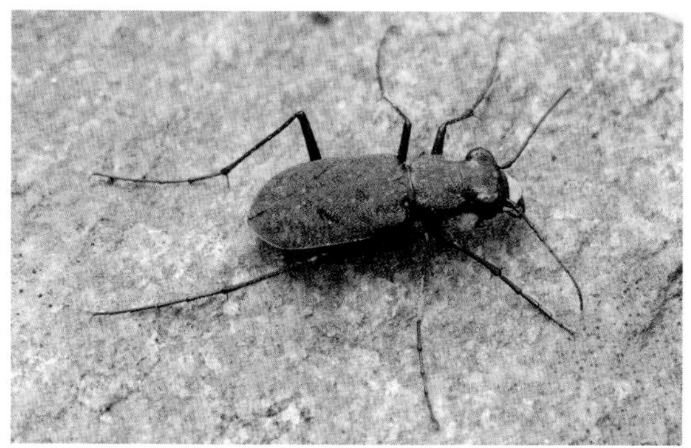

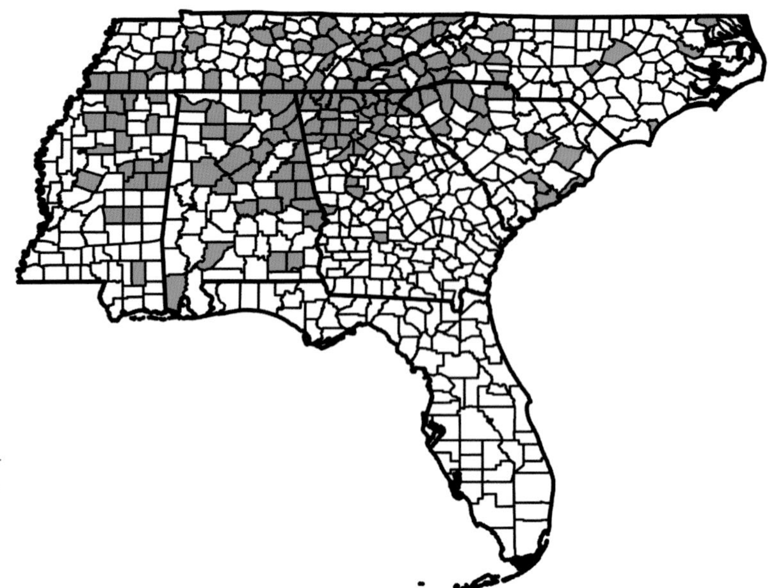

Range of the One-spotted Tiger Beetle (*Apterodela unipunctata*) in the southeastern United States.

Similar Species Distinctive, resembles no other tiger beetle in the region.

Habitat Small openings or dirt paths through forest; also may be found foraging in leaf litter.

Note Inconspicuous and well camouflaged, the One-spotted Tiger Beetle is often seen only when crossing paths or other openings. It is probably the least wary tiger beetle in the region and can often be captured by hand. When disturbed, this species rarely flies, usually preferring to either freeze in place or move slowly and methodically along the ground; any escape flight is short and slow. While fairly common, it is never numerous.

Appalachian Tiger Beetle

Cicindela ancocisconensis T. W. Harris

Size 14–16 mm.

Southeast Status Rare in the Blue Ridge; most records are historical.

Identification A medium-sized brown tiger beetle with mostly complete markings. All dorsal surfaces are a uniform brown to dark brown in color, with somewhat granulate elytra. Often, sutures are slightly edged with metallic green, and the edges of the elytra and area behind the eyes may also have metallic-green highlights. The humeral lunule is thin but complete, and not as curved as some other species; appears more in the shape of the number 7. The middle band is also thin but complete, and the longitudinal segment is often broken. A thin marginal line is present from the middle band running to

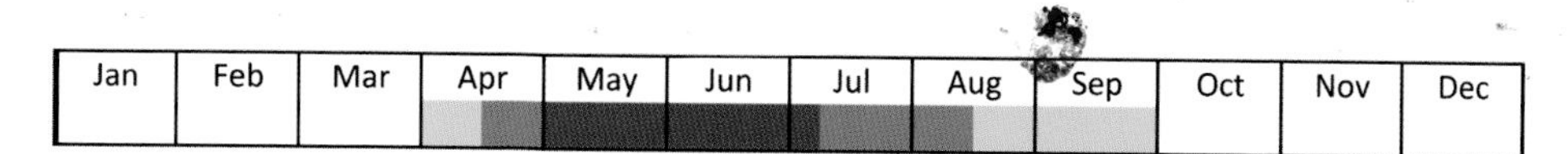

Above: Flight season of the Appalachian Tiger Beetle (*Cicindela ancocisconensis*).

Top to bottom:

Appalachian Tiger Beetle (*Cicindela ancocisconensis*). Scott County, Tennessee.

A strongly marked Appalachian Tiger Beetle (*Cicindela ancocisconensis*). Scott County, Tennessee.

Left to right:

Appalachian Tiger Beetle (*Cicindela ancocisconensis*). Cumberland County, Tennessee.

Appalachian Tiger Beetle (*Cicindela ancocisconensis*). Haywood County, North Carolina.

the posterior but stops short of the apical lunule. The apical lunule is complete. The legs are usually green but sometimes brownish green, and the abdomen is bluish brown.

Similar Species This species is very similar to both the Twelve-spotted Tiger Beetle (*C. duodecimguttata*) and Bronzed Tiger Beetle (*C. repanda*), and may co-occur in the region with either one. The Twelve-spotted Tiger Beetle is fairly easy to distinguish, with all markings usually broken and no marginal line. Darker individuals of the Bronzed Tiger Beetle can be the most difficult to separate. In general, the Appalachian Tiger Beetle is larger; is slightly more slender in shape; has more metallic-green highlights on the head, thorax, or elytra; often has a break in the middle band along the longitudinal segment; often has more setae on the head; and usually has the humeral lunule less curved along its length, shaped more like a 7 (shaped like a J in the Bronzed Tiger Beetle). Some individuals may be very difficult to reliably separate in the field, but in the hand the labrum of the Appalachian Tiger Beetle has three teeth, whereas the Bronzed Tiger Beetle has only one. The Oblique-lined Tiger Beetle (*C. tranquebarica*) is also brown and strongly marked, but the humeral lunule ends in a straight line, there is no marginal line, and this species is usually found away from water.

Habitat Pebbly or sandy shorelines of mountain rivers and streams, or less often in blowouts away from the water.

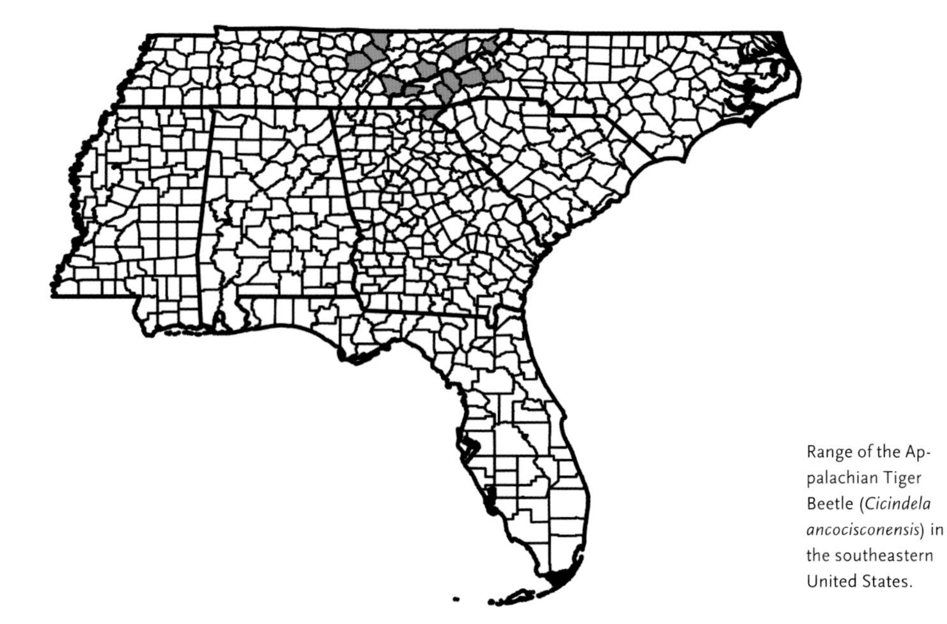

Range of the Appalachian Tiger Beetle (*Cicindela ancocisconensis*) in the southeastern United States.

Note There are few recent records in the Southeast, but this uncommon species occurs immediately to the north; it probably still exists in a few southeastern locations that have not yet been surveyed. At any single location it is likely to be the least numerous species. The fall season may be very brief.

Twelve-spotted Tiger Beetle

Cicindela duodecimguttata Dejean

Size 12–15 mm.

Southeast Status Fairly common in the northern part of the region, becoming rarer to the south. Absent from most of the Coastal Plain and Florida.

Identification A medium-sized beetle that is mostly dark brown, with a variable set of pale markings often broken into twelve discrete spots or marks. The head and pronotum are dark brown, with some green highlights along the sutures or edges. The labrum is white. The pronotum is asymmetrical, wider at the anterior end than the posterior when viewed dorsally. Elytra are dark brown and somewhat granulate. The humeral lunule is typically broken into two spots, but the spot at the anterior end may be absent or, rarely, the full humeral lunule may be present. The middle band is usually broken into a complete transverse segment and a discrete posterior medial spot. Less commonly the transverse segment is thin or broken, and rarely the entire middle band is present. The marginal line is usually absent.

Jan	Feb	Mar	Apr	May	Jun	Jul	Aug	Sep	Oct	Nov	Dec

Above: Flight season of the Twelve-spotted Tiger Beetle (*Cicindela duodecimguttata*).

Right: Twelve-spotted Tiger Beetle (*Cicindela duodecimguttata*). Lowndes County, Mississippi.

The apical lunule is broken into two discrete spots, with spot along elytral apices more elongate. The legs are brown with some metallic-green areas. The abdomen is dark blue to black.

Similar Species Most often confused with the Bronzed Tiger Beetle (*C. repanda*), but also similar to the Appalachian Tiger Beetle (*C. ancocisconensis*). The Bronzed Tiger Beetle usually has paler brown elytra color with more complete and often thicker markings, including complete humeral and apical lunules. The marginal line is present and nearly complete. The pronotum of the Bronzed Tiger Beetle is symmetrical. Can also be confused with the Appalachian Tiger Beetle in the region, and the two may co-occur along mountain streams and rivers. Both have dark brown elytra, but the Appalachian Tiger Beetle has complete apical and humeral lunules, and the marginal line is present posteriorly from the transverse segment of the middle band. The Appalachian Tiger Beetle also has stronger green highlights at sutures.

Top to bottom:

Twelve-spotted Tiger Beetle (*Cicindela duodecimguttata*). Lowndes County, Mississippi.

Twelve-spotted Tiger Beetle (*Cicindela duodecimguttata*). Putnam County, Tennessee.

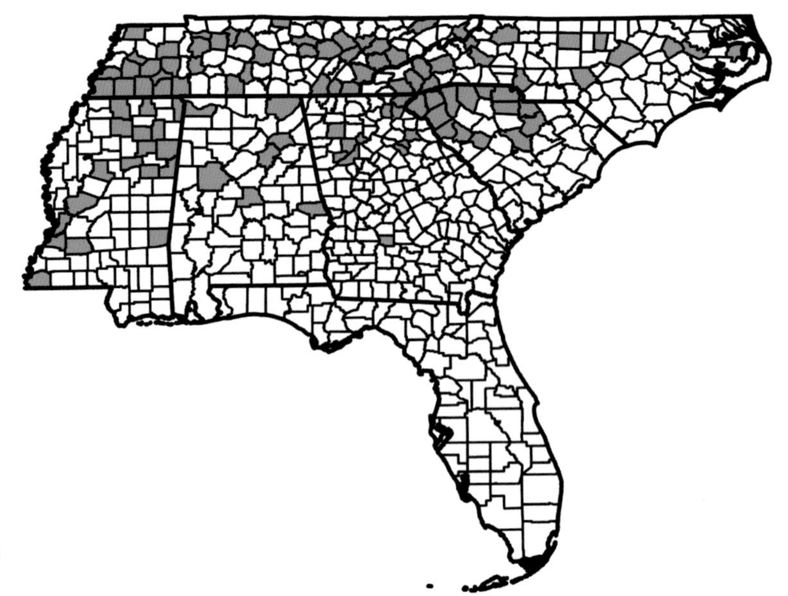

Range of the Twelve-spotted Tiger Beetle (*Cicindela duodecimguttata*) in the southeastern United States.

Habitat Found in diverse habitats, but usually along the edges of ponds, lakes, streams, and rivers. Prefers silty rather than sandy soils, but also often found among rocks and pebbles.

Note Can occur in numbers in appropriate habitat. Has declined significantly during the last century in the Piedmont of the Carolinas and Georgia based on large numbers in museum collections and scarcity today. Most consistently found along rivers or large streams in or near the Blue Ridge and along small to medium streams in Mississippi. Interestingly, in Tennessee, also found reliably around ponds and lakes.

Big Sand Tiger Beetle

Cicindela formosa generosa Dejean

Size 16–21 mm.

Southeast Status Uncommon in Mississippi and western Tennessee; barely reaches western Alabama. Absent from remainder of region.

Identification Our largest Temperate Tiger Beetle, this is a robust brown beetle with wide obvious maculations. The head is brown with a white labrum. The face is often covered with setae. The pronotum is also brown, and the sides of the thorax are often thickly covered with

Jan	Feb	Mar	Apr	May	Jun	Jul	Aug	Sep	Oct	Nov	Dec

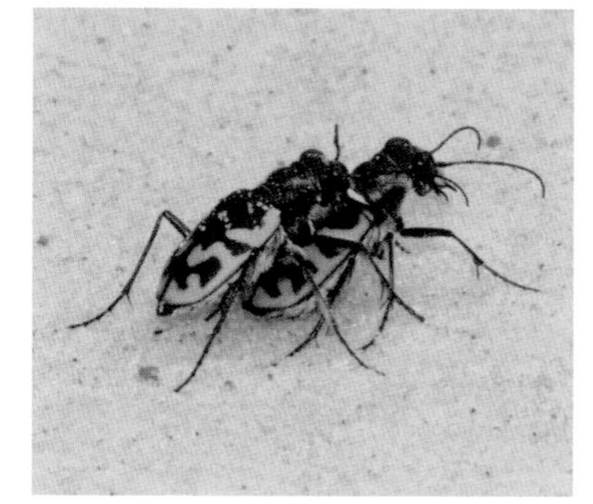

Above: Flight season of the Big Sand Tiger Beetle (*Cicindela formosa generosa*).

Clockwise from top left:

Big Sand Tiger Beetle (*Cicindela formosa generosa*). Stone County, Mississippi.

Big Sand Tiger Beetle (*Cicindela formosa generosa*), Stone County, Mississippi.

Pair of Big Sand Tiger Beetles (*Cicindela formosa generosa*). Stone County, Mississippi.

Big Sand Tiger Beetle (*Cicindela formosa generosa*). Mobile County, Alabama.

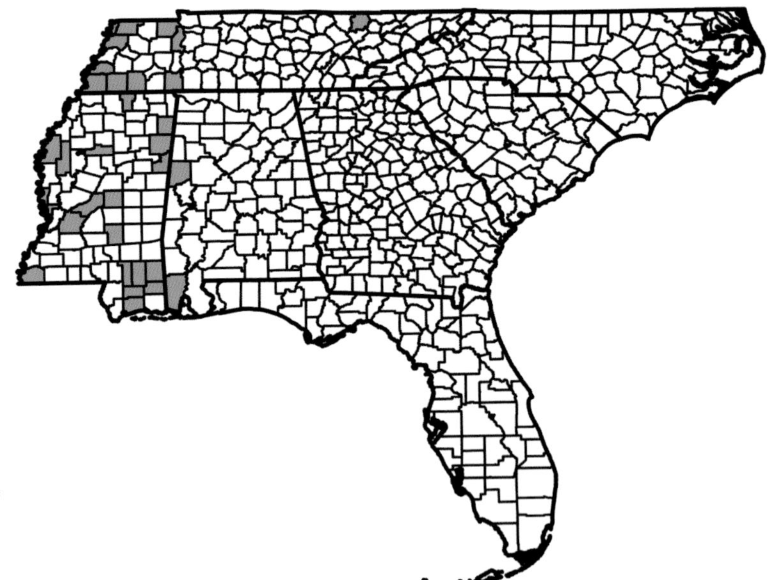

Range of the Big Sand Tiger Beetle (*Cicindela formosa generosa*) in the southeastern United States.

setae. The elytral color is usually dark brown; maculations are usually white but may be cream colored. All maculations are complete and bold, with a complete, expanded marginal line forming a wide band that connects all the marks. The posterior end of the humeral lunule is angled and sometimes quite short. The legs are usually brown with metallic green highlights, although on some individuals they are mostly green. The abdomen is dark green or black.

Similar Species No other tiger beetles in the region share the large size and maculations of the Big Sand Tiger Beetle. The Oblique-lined Tiger Beetle (*C. tranquebarica*) has a similar shape to the humeral lunule, but it is a smaller beetle with much thinner markings.

Habitat Deep, soft, dry sand of large river bars and open areas.

Note Appears later in the morning than other tiger beetles due to its large size. Usually makes very long escape flights, often rolling on landing, and then frequently runs rapidly through sparse vegetation. Very often seen consuming ants.

Hairy-necked Tiger Beetle

Cicindela hirticollis hirticollis Say

Size 10–15 mm.

Southeast Status Uncommon to rare in two very different habitats: along Atlantic coastal beaches, and interior river sandbars primarily in Mississippi, Alabama and Tennessee.

Identification A medium-sized tiger beetle that varies in overall color from dark brown with wide maculations (inland form) to grayish blue or green with thinner maculations (coastal form). The coastal form is smaller than the inland form. The following applies to both forms: The head and pronotum have numerous setae. The humeral lunule is complete and curves at the posterior end to form a mark

Jan	Feb	Mar	Apr	May	Jun	Jul	Aug	Sep	Oct	Nov	Dec

Above: Flight season of the Hairy-necked Tiger Beetle (*Cicindela hirticollis hirticollis*).

Top to bottom:

Inland form Hairy-necked Tiger Beetle (*Cicindela hirticollis hirticollis*). Perry County, Alabama.

Pair of inland form Hairy-necked Tiger Beetles (*Cicindela hirticollis hirticollis*). Perry County, Alabama.

Left to right:

Coastal form Hairy-necked Tiger Beetle (*Cicindela hirticollis hirticollis*) digging refugium. Glynn County, Georgia.

Coastal form Hairy-necked Tiger Beetle (*Cicindela hirticollis hirticollis*). Glynn County, Georgia.

in the shape of a capital G. The middle band is complete; the marginal line extends forward to connect with humeral lunule and extends rearward almost to apical lunule. The apical lunule is complete. The abdomen is metallic blue or green. The legs are brown or coppery with some green highlights; the green is more prominent in the coastal form.

Similar Species Most easily confused with the highly variable Bronzed Tiger Beetle (*C. repanda*). On coastal beaches, it's usually not a problem to distinguish the Hairy-necked Tiger Beetle, but along inland rivers the two species can be quite similar. The Bronzed Tiger Beetle is usually smaller in size and lighter brown in color with thinner maculations, but in the case of strongly marked individuals or a lightly marked Hairy-necked Tiger Beetle, the shape of the humeral lunule is key. In the Bronzed Tiger Beetle the humeral lunule is in the shape of a capital J; very rarely does the posterior end curve around at all, and even then the curve is not as pronounced as in typical individuals of the Hairy-necked Tiger Beetle. As the name implies, the Hairy-necked Tiger Beetle is often much more densely covered with setae on the thorax than the Bronzed Tiger Beetle. The Bronzed Tiger Beetle vastly outnumbers the Hairy-necked Tiger Beetle almost everywhere the two species co-occur.

Habitat Coastal forms are found only on less populated beaches, and are most often near some kind of water on the upper beach such as ponds, large depressions, or creeks flowing through the beach to get

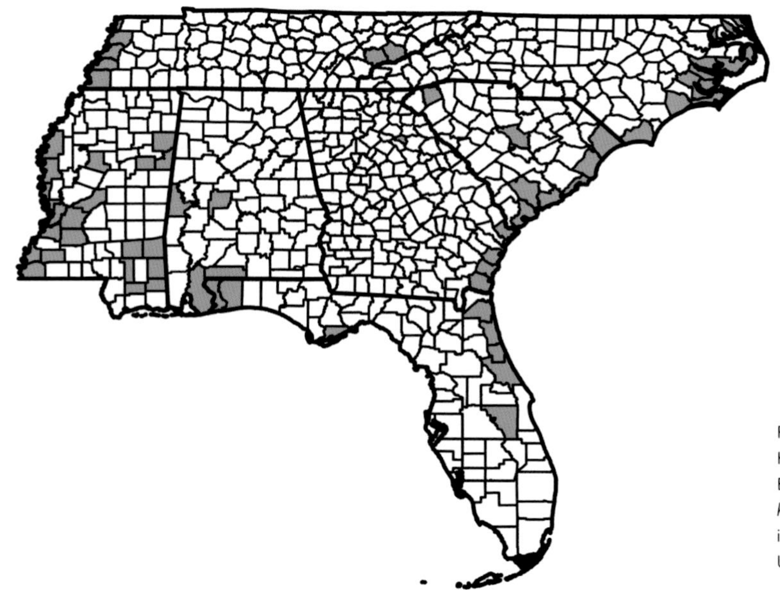

Range of the Hairy-necked Tiger Beetle (*Cicindela hirticollis hirticollis*) in the southeastern United States.

to the ocean. The more common inland form is found on wetter portions of large river bars, and individuals usually occur on muddier, rather than sandier, sections of the bars.

Note Both forms are quite wary. The coastal form often roosts communally in sides of dunes just above the upper beach. This form is rare and declining in the Southeast.

Autumn Tiger Beetle

Cicindela nigrior Schaupp

Size 11–14 mm.

Southeast Status Scattered and local along Fall Line and in Coastal Plain from North Carolina to Mississippi; can be locally numerous. Only active late summer to early winter.

Identification A medium sized green, blue-green, or black tiger beetle, always totally devoid of markings. The green or blue-green form occurs in far eastern populations, and the black form in far western populations. Both forms may be found in an overlap zone that roughly follows a line from Fort Walton Beach, Florida, northeast through Augusta, Georgia. Regardless of form, all surfaces of this beetle, including the legs, are the same color. A few setae are scattered on the head. The labrum is black in females and black with two symmetrically placed white spots resembling headlights in males. The labrum has three teeth in both sexes, with the blunt middle tooth shorter than

Jan	Feb	Mar	Apr	May	Jun	Jul	Aug	Sep	Oct	Nov	Dec

Above: Flight season of the Autumn Tiger Beetle (*Cicindela nigrior*).

Right: Labrum of a female green form Autumn Tiger Beetle (*Cicindela nigrior*). Aiken County, South Carolina.

the outer teeth. The elytral apices do not have an upturned marginal edge. The abdomen is dark blue to black.

Similar Species No other tiger beetle in the region is black without any markings. Green forms may be found very near green Festive Tiger Beetles (*C. scutellaris unicolor*) as they both occur in very similar sandy habitat, but many examples of the Festive Tiger Beetle found outside Florida have at least a small white mark for an apical lunule, and males have a solid white labrum. Females of both species have a black labrum, but the teeth are different. In the Festive Tiger Beetle the sharply pointed middle tooth is longer than the outer teeth. Green or blue-green examples of the Festive Tiger Beetle are shinier than Autumn Tiger Beetles and slightly less robust in appearance. Festive Tiger Beetle males also generally have more setae on the head. Additionally, the apical elytral margins (where the apical lunules are, if present) of both sexes are upturned, forming a small lip. This character is subtle and may only be seen upon very close inspection. Lastly, the Autumn Tiger Beetle is only active in fall, and breeds then. The Festive Tiger Beetle is active in spring and fall but only breeds in spring, so any pairs found in fall should be Autumn Tiger Beetles; this can be easily confirmed by examining the male labrum.

Clockwise from top left:

Green form Autumn Tiger Beetle (*Cicindela nigrior*). Emanuel County, Georgia.

Green form Autumn Tiger Beetle (*Cicindela nigrior*). Emanuel County, Georgia.

Pair of green form Autumn Tiger Beetles (*Cicindela nigrior*). Aiken County, South Carolina.

Pair of black form Autumn Tiger Beetles (*Cicindela nigrior*). Okaloosa County, Florida.

Black form Autumn Tiger Beetle (*Cicindela nigrior*). Taylor County, Georgia.

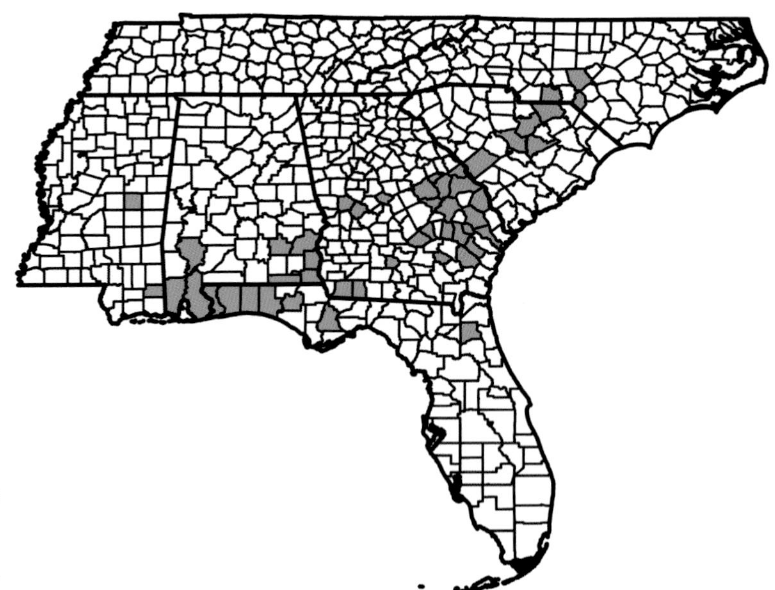

Range of the Autumn Tiger Beetle (*Cicindela nigrior*) in the southeastern United States.

Unmarked individuals of the Six-spotted Tiger Beetle (*C. sexguttata*) are somewhat similar to the green or blue-green form Autumn Tiger Beetle but have granulate elytra; both sexes have a white labrum; and adults are active earlier in the year and in very different habitats.

Habitat Typically found in well-packed dry sand, with sparse to moderate grassy vegetation. Autumn Tiger Beetles are often found running through areas of moderate vegetation. Festive Tiger Beetles, by contrast, prefer deeper loose sand and sparse vegetation.

Note Often prefers to be in or near vegetation, and when flushed, often lands right next to the base of vegetation and sits motionless. On cool mornings the Autumn Tiger Beetle may be slow to emerge and prefers to run rather than fly until temperatures warm up. This species was formerly considered a subspecies of the Festive Tiger Beetle (Vick and Roman 1985).

Northern Barrens Tiger Beetle

Cicindela patruela patruela Dejean

Size 12–14 mm.

Southeast Status Extremely rare; probably restricted to a few sites in or near the Blue Ridge.

Identification A medium-sized green tiger beetle with white markings. The head, pronotum, and elytra are a uniform velvety green color, which is less shiny than in similar species. The labrum is white. The humeral lunule is always present but is reduced to two discrete spots. The middle band is usually complete and consists of only the transverse segment, expanded at both ends. There is no marginal line. The apical lunule is also reduced to two discrete spots. The legs are green, and the abdomen is dark green or black.

Similar Species The Six-spotted Tiger Beetle (*C. sexguttata*) is very similar and often found in the same habitats during spring, but it has no humeral lunule markings and is much shinier green. The Six-spotted Tiger Beetle virtually always has the middle band reduced to discrete spots. The green form Festive Tiger Beetle (*C. scutellaris unicolor*) is also shinier green with very different and/or reduced markings, and the species is not found in the Southeast at the elevations associated with the Northern Barrens Tiger Beetle.

Below: Flight season of the Northern Barrens Tiger Beetle (*Cicindela patruela patruela*).

Jan	Feb	Mar	Apr	May	Jun	Jul	Aug	Sep	Oct	Nov	Dec

Left to right:

Northern Barrens Tiger Beetle (*Cicindela patruela patruela*). Wolfe County, Kentucky.

Northern Barrens Tiger Beetle (*Cicindela patruela patruela*). Pike County, Ohio.

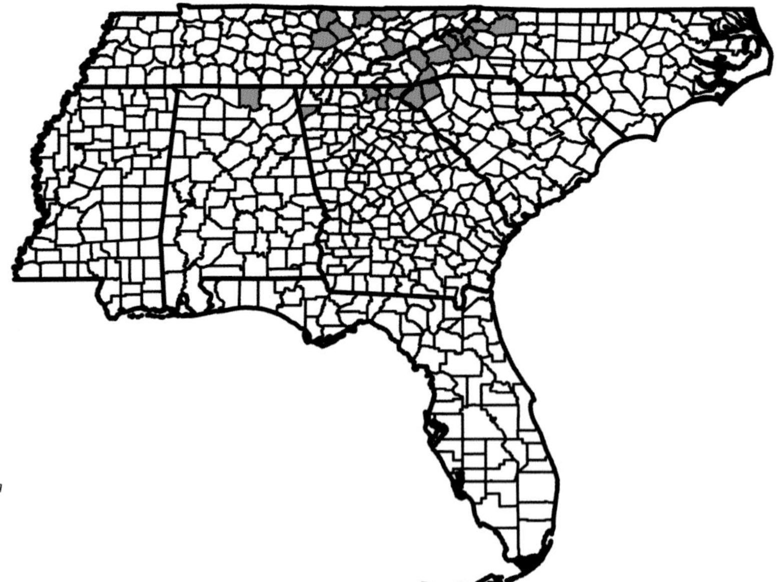

Range of the Northern Barrens Tiger Beetle (*Cicindela patruela patruela*) in the southeastern United States.

Habitat Found along trails or other forest openings, especially on dry pebbly dirt or shallow sandstone hillsides. Often found where lichens and mosses are present, but not found in leaf litter.

Note Known historically from several locations within the southeastern Blue Ridge and Appalachian Plateau, including one locality in extreme north Alabama; currently known in the Southeast only from near the Blue Ridge Parkway in North Carolina. Has not been seen anywhere else in the region in over 50 years; decline may be related to fire suppression, as this species often occurs in recently burned areas. The Northern Barrens Tiger Beetle is easy to spot while actively foraging, but escape flights are typically into vegetation, where it is difficult to relocate.

Cow Path Tiger Beetle

Cicindela purpurea purpurea Olivier

Size 12–15 mm.

Southeast Status Rare and local; restricted to higher portions of the Blue Ridge or scattered areas in Tennessee and north Mississippi.

Identification A medium-sized mostly reddish-purple or brown tiger beetle with reduced markings. The general ground color of this species can vary from reddish purple to brown to almost green. The head, pronotum, and elytra are uniform in color, and the margins of the thorax and elytra, including sutures, are usually metallic green. The labrum is white. The humeral lunule is usually absent, but one or two small spots may occasionally be present. The middle band is short and reduced to a single wavy line without expansions at either end; this line does not extend to the edge of the elytra. There is no marginal line. The apical lunule is usually reduced to two small marks, with the mark along the elytral apices reduced to a short thin line. The legs are reddish brown, with some green highlights in greener individuals. The abdomen is metallic blue to green.

Below: Flight season of the Cow Path Tiger Beetle (*Cicindela purpurea purpurea*).

Clockwise from top left:

Cow Path Tiger Beetle (*Cicindela purpurea purpurea*). Humphreys County, Tennessee.

Cow Path Tiger Beetle (*Cicindela purpurea purpurea*). Humphreys County, Tennessee.

Cow Path Tiger Beetle (*Cicindela purpurea purpurea*). Haywood County, North Carolina.

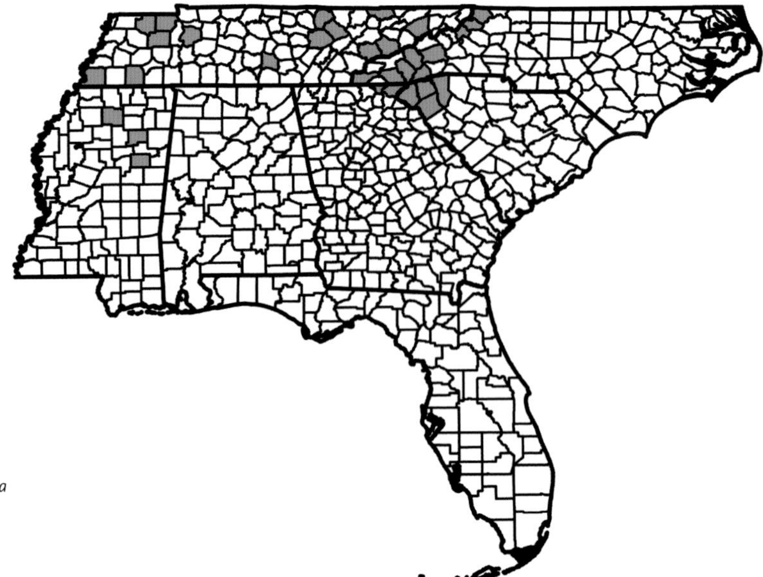

Range of the Cow Path Tiger Beetle (*Cicindela purpurea purpurea*) in the southeastern United States.

Similar Species The only similar species in the region is the Splendid Tiger Beetle (*C. splendida*), which is much more common in similar locations and habitats. The Splendid Tiger Beetle always has a green head and thorax, contrasting with reddish-brown elytra, while the Cow Path Tiger Beetle is always uniform in color.

Habitat Higher elevation open bare ground, with soil types ranging from clay to silt.

Note This is another species that seems to have declined considerably in the Southeast, with few recent observations. It is very wary, but usually makes only short escape flights and lands facing the perceived threat (which is usually the observer).

Bronzed Tiger Beetle

Cicindela repanda repanda Dejean

Size 10–13 mm.

Southeast Status Common to abundant throughout the region, but absent from peninsular Florida.

Identification A variable medium-sized brownish tiger beetle. All dorsal surfaces and legs are uniform in color, but that color ranges from flat dark brown to metallic dark brown to a considerably paler metallic bronzy brown. Examples of all of these may occur in the same population. The labrum, small and white in both sexes, has a single tooth. The thorax is often covered on the sides with setae. The elytra are strongly granulate—this feature can be seen with binoculars. Typical elytral markings are as follows: The humeral lunule is complete and in the shape of a capital J or shallow C. The middle

Below: Flight season of the Bronzed Tiger Beetle (*Cicindela repanda repanda*).

Clockwise from top left:

Bronzed Tiger Beetle (*Cicindela repanda repanda*). Perry County, Alabama.

Bronzed Tiger Beetle (*Cicindela repanda repanda*). Upson County, Georgia.

Bronzed Tiger Beetle (*Cicindela repanda repanda*) with prey. Grady County, Georgia.

Strongly marked Bronzed Tiger Beetle (*Cicindela repanda repanda*). Oconee County, South Carolina.

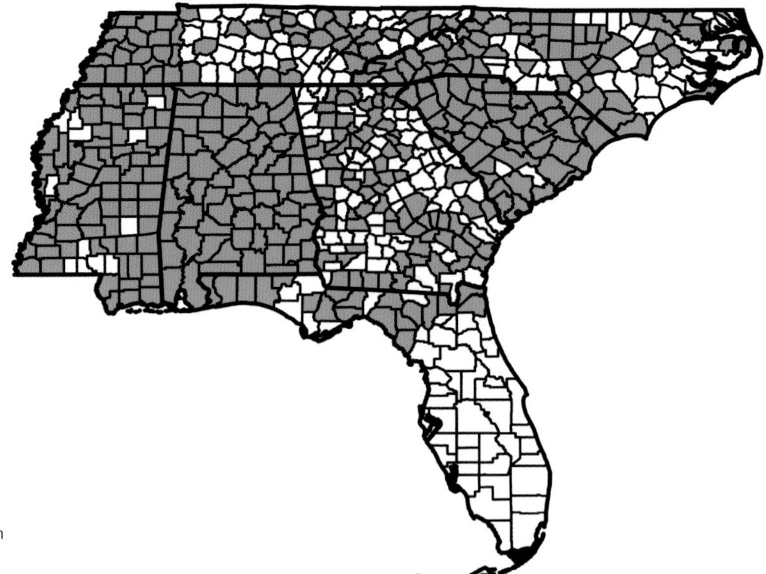

Range of the Bronzed Tiger Beetle (*Cicindela repanda repanda*) in the southeastern United States.

band is complete and usually terminates in an expanded spot. The marginal line is present and may extend nearly to both the humeral and apical lunules. The apical lunule is complete. In a typical Bronzed Tiger Beetle, all these marks are complete and thin. However, in some individuals the marks may be much thicker, and in other individuals the marks may be so thin as to be incomplete. The humeral lunule in particular is variable, with the shape ranging from the typical J, to the posterior end curving around almost into a fully rounded C or, more rarely, ending in a straight line. The metallic-green abdomen is often seen protruding past the tips of the elytra.

Similar Species This species can resemble many other well-marked brown tiger beetles due to its variability. Typical bronzed individuals are uniquely colored, but darker brown individuals can resemble the inland Hairy-necked Tiger Beetle (*C. hirticollis*), Oblique-lined Tiger Beetle (*C. tranquebarica*), Twelve-spotted Tiger Beetle (*C. duodecimguttata*), or Appalachian Tiger Beetle (*C. ancocisconensis*). In all cases the elytra of the Bronzed Tiger Beetle are more granular, and much of the time they are also more reflective or metallic in appearance. Easiest to eliminate is the Oblique-lined Tiger Beetle; it is not found near water, and the posterior portion of the humeral lunule of the Bronzed Tiger Beetle is almost never as straight as on the Oblique-lined Tiger Beetle.

The Hairy-necked Tiger Beetle can be separated by the combination of very thick markings and the humeral lunule being curved all the way around into a G. Though the Bronzed Tiger Beetle can have dense setae on the sides of the thorax, the setae are not as dense as those found on the Hairy-necked Tiger Beetle. In the Twelve-spotted Tiger Beetle the markings are almost always broken, the asymmetrical thorax has the anterior end wider than the posterior while the thorax of the Bronzed Tiger Beetle is symmetrical, and the marginal line is absent. The Appalachian Tiger Beetle is the most difficult to separate and may co-occur with Bronzed Tiger Beetles at some sites. In general, the Appalachian Tiger Beetle is larger and slightly more slender in shape; has more metallic-green highlights on the head, pronotum, or elytra; often has a broken middle band along the longitudinal segment; often has more setae on the head; and usually has less curvature along the length of the humeral lunule, more like a 7 than a J. Some individuals may be very difficult to separate in the field, but in the hand the labrum of the Appalachian Tiger Beetle has three teeth while the Bronzed Tiger Beetle has only one.

Habitat Common to abundant in sandy areas along any type of running water, from small streams to the largest rivers. Found on tidal rivers downstream almost to salt water. Occasionally found in other habitats, such as around lakes and ponds or in dry sandy areas near but not necessarily adjacent to rivers.

Note During peak season, sandbars along rivers may be literally covered with hundreds or thousands of this abundant species. Unlike most spring/fall species, scattered individuals of this incredibly common tiger beetle can usually be found throughout the summer. Often seen scavenging on carrion and dead insects.

Festive Tiger Beetle

Cicindela scutellaris Say

Size 11–14 mm.

Southeast Status Fairly common in appropriate habitat throughout much of the region, but largely absent from most of Alabama, northern Georgia, eastern and middle Tennessee, and southern Florida.

Identification This medium-sized tiger beetle is extremely variable, with four subspecies or subspecies intergrades occurring in the Southeast. In all regional subspecies or forms the labrum is white in males and black in females. In both sexes the sharply pointed middle tooth of the labrum is longer than those to either side.

By far the most widespread is *C. s. unicolor* Dejean, which is uniformly shiny green or blue green and mostly unmarked. All body parts including legs are entirely green or blue green, with setae on the dorsal surface of the head. Setae are more pronounced in males.

Below: Flight season of the Festive Tiger Beetle (*Cicindela scutellaris* sspp.).

Jan	Feb	Mar	Apr	May	Jun	Jul	Aug	Sep	Oct	Nov	Dec

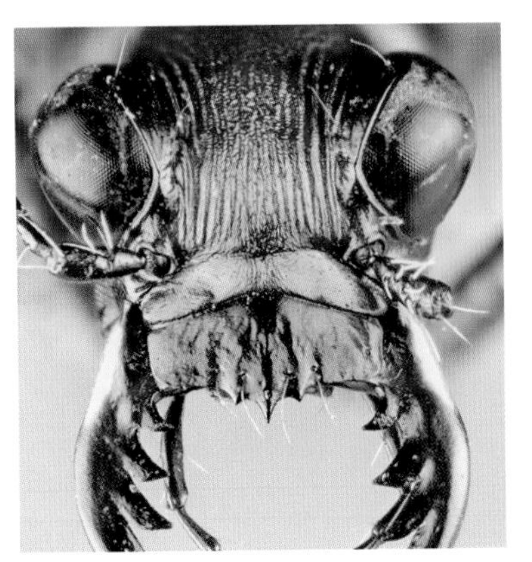

Clockwise from top left:

Labrum of a female Festive Tiger Beetle (*Cicindela scutellaris unicolor*). Pasco County, Florida.

Festive Tiger Beetle (*Cicindela scutellaris unicolor*). Pike County, Alabama.

Festive Tiger Beetle (*Cicindela scutellaris unicolor*). Pasco County, Florida.

Very rarely, and usually in areas within the range of the Autumn Tiger Beetle (*C. nigrior*), the male labrum has a thin black border. The only elytral marking in this subspecies is a small, straight white mark often found at the apices; this mark is the medial remnant of the apical lunule. The posterior edge of the elytra has a lip where this mark is found. The abdomen is metallic green.

The subspecies *C. s. rugifrons* Dejean is also uniformly green, but with expanded white elytral markings. The humeral lunule is usually absent, but rarely a tiny spot is present at the posterior end of this mark. The middle band consists of a relatively large discrete triangular mark on the margin of the elytra; this mark does not extend transversely into the interior of the elytra except for the point of the triangle. There is usually no marginal line, although in some individuals the middle band and apical lunule expand enough to connect. The apical lunule consists of either two discrete marks or one larger continuous apical lunule. All of the above elytral marks are often edged with a very thin dark border. The legs and abdomen are green. This subspecies occurs mostly to the north along the Atlantic seaboard; rare intergrades between *C. s. rugifrons* and *C. s. unicolor* are found in the Carolinas. There are a few specimen records in the region (mostly from North Carolina) showing almost full *C. s. rugifrons* markings.

Clockwise from top left:

Festive Tiger Beetle (*Cicindela scutellaris unicolor*). Emanuel County, Georgia.

Festive Tiger Beetle (*Cicindela scutellaris unicolor*). Emanuel County, Georgia.

Festive Tiger Beetle (*Cicindela scutellaris unicolor*). Emanuel County, Georgia.

The subspecies *C. s. lecontei* Haldeman is either reddish brown or green, but is highly variable with respect to both color and white elytral markings. Some individuals have a greenish head and pronotum but brown or dull brownish-green elytra. The humeral lunule is usually present, but may be either a spot at the posterior end of this mark or a thick complete line following the margin of the elytra. The middle band consists of a discrete triangular mark on the edge of the elytra; the mark does not extend into the interior of the elytra. This triangle may be quite small, or it may be flattened to extend along the length of the margin of the elytra. The apical lunule is usually one larger continuous marking of variable thickness and extending almost to the middle band. The legs are brown with some green highlights. This subspecies occurs mostly in the Northeast and Midwest of the US; in the Southeast, intergrades between it and *C. s. unicolor* have been found mostly in northeast Mississippi and southwest Tennessee, with a few individuals showing almost full *C. s. lecontei* markings.

The subspecies *C. s. scutellaris* Say is a beautiful beetle with a metallic-green head and thorax, and the green color extends onto the interior of the elytra and blends into a vivid reddish orange on the posterior portion of the elytra. In the Southeast, this subspecies occurs only as an intergrade with others (chiefly *C. s. lecontei*). In these intergrades, the colors are as described above, with white markings

Clockwise from top left:

Festive Tiger Beetle (*Cicindela scutellaris rugifrons*). Newport County, Rhode Island.

Festive Tiger Beetle (*Cicindela scutellaris lecontei*). Tishomingo County, Mississippi.

Festive Tiger Beetle (*Cicindela scutellaris lecontei*). Tishomingo County, Mississippi.

either absent or reduced as follows: The humeral lunule is usually absent, and the middle band consists only of a small discrete spot on the edge of the elytra. The apical lunule usually consists of a continuous complete mark. In some individuals the middle band and apical lunule expand enough to barely connect along the marginal line. The legs and abdomen are green. Intergrades of *C. s. scutellaris* are found mostly along the Mississippi River in extreme western Mississippi.

Similar Species All subspecies of the Festive Tiger Beetle in the region are fairly unique except for *C. s. unicolor*, which is similar to both the green form Autumn Tiger Beetle and Six-spotted Tiger Beetle (*C. sexguttata*); all comments in this section pertain to *C. s. unicolor*. Many Festive Tiger Beetles north of Florida have a small white marking at the elytral apices, but no other marks; the Autumn Tiger Beetle is always completely unmarked. In Festive Tiger Beetle males the labrum is white, but in Autumn Tiger Beetle males the labrum is black with two white spots; this feature can be seen through binoculars. Females of both species have a black labrum. In both sexes the middle tooth on the labrum can be used to separate the species when viewed under magnification, either in the field with a loupe or under a microscope. In the Festive Tiger Beetle the sharply pointed middle tooth is longer than the adjacent teeth; in the Autumn Tiger Beetle

Clockwise from top left:

Festive Tiger Beetle (*Cicindela scutellaris lecontei*). Hartford County, Connecticut.

Festive Tiger Beetle (*Cicindela scutellaris scutellaris*). Bolivar County, Mississippi.

Festive Tiger Beetle (*Cicindela scutellaris scutellaris*). Bolivar County, Mississippi.

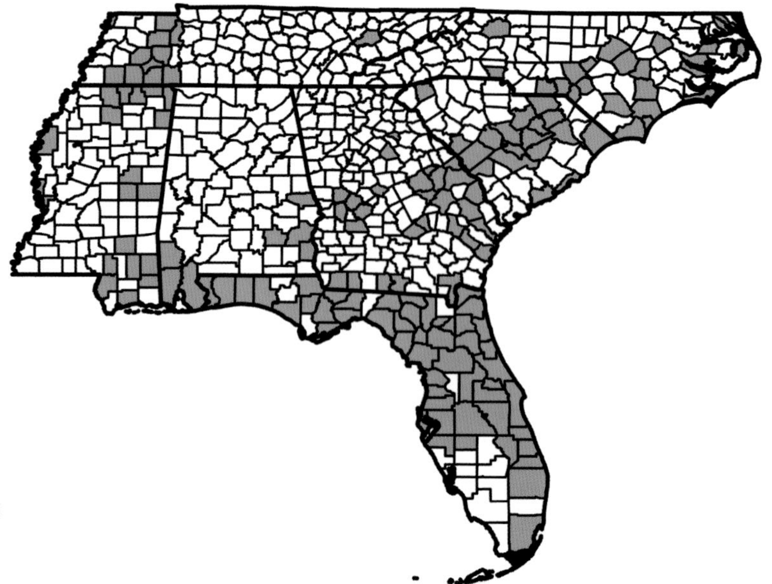

Range of the Festive Tiger Beetle (*Cicindela scutellaris* sspp.) in the southeastern United States.

the rounded middle tooth is shorter than the adjacent teeth. The Autumn Tiger Beetle is less shiny than the Festive Tiger Beetle and is slightly more robust in appearance. The Autumn Tiger Beetle is also only active in the fall. Most Six-spotted Tiger Beetles are more heavily marked than the Festive Tiger Beetle. Unspotted Six-spotted Tiger Beetles can be differentiated from unmarked Festive Tiger Beetles by coarser elytra, a feature that is visible through binoculars; slight differences in color as the Festive Tiger Beetle is darker green or blue-green; more slender shape; lack of setae on head; and the white labrum of females.

Habitat All subspecies and intergrades are found in loose dry sand, usually with little vegetation. Open fields and paths or fire roads through sandy forests are suitable habitats for this species. Sometimes just a few open patches in an otherwise well-vegetated area will provide enough habitat to support a population.

Note This species is less wary than many tiger beetles, and it usually makes short escape flights into open areas (not landing near vegetation). Often active throughout the day, but seeks shade during hotter periods.

Six-spotted Tiger Beetle

Cicindela sexguttata Fabricius

Size 10–14 mm.

Southeast Status Common throughout much of region; absent from most of peninsular Florida.

Identification This medium-sized tiger beetle is a shiny emerald green (less commonly blue green), with markings reduced to a varying number of small white spots. The head is green; the labrum is white in both sexes. The pronotum is wider than long. The elytra are green, and the surface is granulate and not smooth in appearance. The elytra typically have six spots, but strongly marked individuals may have eight spots, and poorly marked individuals (seen more often in the southern part of the range) may have fewer or no spots. The humeral lunule is absent. The middle band typically has spots only

Jan	Feb	Mar	Apr	May	Jun	Jul	Aug	Sep	Oct	Nov	Dec

Above: Flight season of the Six-spotted Tiger Beetle (*Cicindela sexguttata*).

Top to bottom:

Six-spotted Tiger Beetle (*Cicindela sexguttata*). Bladen County, North Carolina.

Six-spotted Tiger Beetle (*Cicindela sexguttata*). Perry County, Alabama.

Clockwise from top left:

Six-spotted Tiger Beetle (*Cicindela sexguttata*). Perry County, Alabama.

Six-spotted Tiger Beetle (*Cicindela sexguttata*). Morgan County, Tennessee.

Six-spotted Tiger Beetle (*Cicindela sexguttata*). Rabun County, Georgia.

at the outside edges but occasionally there are additional spots at its medial end; in rare cases there is a nearly complete thin middle band. The marginal line is absent. The apical lunule is always incomplete or absent; it is typically reduced to two spots representing each end of the apical lunule, and the inner spot near the suture is usually larger. The legs and abdomen are green.

Similar Species The Northern Barrens Tiger Beetle (*C. patruela*) is a different shade of green; the elytra are not clearly granulate but more velvet-like in texture, with much more extensive markings always including a complete humeral lunule and a thick, usually complete middle band. The Festive Tiger Beetle (*C. scutellaris* sspp.) has much smoother and shinier elytra, and all marked green subspecies have different patterns of spots and markings than the Six-spotted Tiger Beetle. Unspotted Six-spotted Tiger Beetles can be differentiated from unmarked Festive Tiger Beetles (*C. s. unicolor*) by granulate elytra, which are visible through binoculars; slight differences in color as the Festive Tiger Beetle is darker green or blue green; more slender shape; lack of setae on head; and the white labrum of females—female Festive Tiger Beetles have black labrum. Unspotted Six-spotted

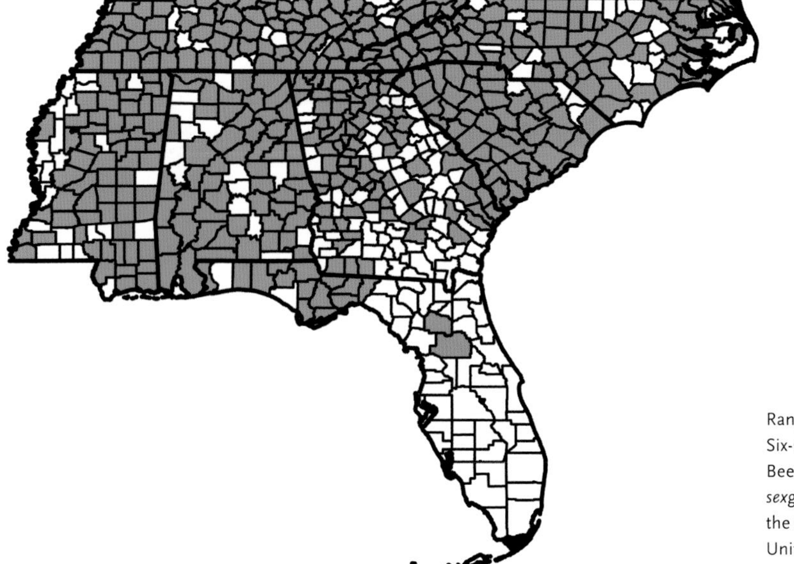

Range of the Six-spotted Tiger Beetle (*Cicindela sexguttata*) in the southeastern United States.

Tiger Beetles can be separated from green form Autumn Tiger Beetles (*C. nigrior*) by the same features as the Festive Tiger Beetle plus differences in the male labrum color pattern. The Six-spotted Tiger Beetle is usually nearly completely separated from the Autumn Tiger Beetle by range, habitat, and flight season.

Habitat Deciduous forest clearings and paths, usually in sunny gaps. Often seen foraging along downed tree trunks. Can be found along forested streams and rivers in lower numbers, sometimes on rocks along the shore. Also occurs in lower densities in fields and other open areas near forests.

Note Commonly found along trails, this beetle is often noticed by hikers due to its bright color and habit of flushing repeatedly. Will typically fly short escape flights to the next sunny spot. Sometimes flies into shade or vegetation to escape. Very few individuals are found in fall. These specimens are usually recently metamorphosed adults that will overwinter after a short period of emergence; most such adults remain in their underground pupal chambers until the following spring (Knisley and Schultz 1997). Adults are known to hibernate under tree bark, and individuals can occasionally be encountered in early spring on unseasonably warm days.

Splendid Tiger Beetle

Cicindela splendida Hentz

Size 12–15 mm.

Southeast Status Uncommon to rare in or near the Blue Ridge, western Tennessee, and northern Mississippi. Possibly declining, as historical records are more numerous and somewhat more wide-ranging than current known records indicate.

Identification A gorgeous medium-sized green and reddish-brown tiger beetle. The head, thorax, and legs are bright metallic green. The labrum is white. The elytra are usually bright reddish brown but may be a brighter reddish purple or, rarely, reddish green. The margins of the elytra, and often the elytral suture, are edged with bright metallic green. White elytral markings are minimal. The humeral lunule is typically absent, but if present, it consists of a single small spot at

Below: Flight season of the Splendid Tiger Beetle (*Cicindela splendida*).

Clockwise from top left:

Splendid Tiger Beetle (*Cicindela splendida*). Towns County, Georgia.

Splendid Tiger Beetle (*Cicindela splendida*). Towns County, Georgia.

Splendid Tiger Beetle (*Cicindela splendida*). Towns County, Georgia.

Splendid Tiger Beetle (*Cicindela splendida*). Towns County, Georgia.

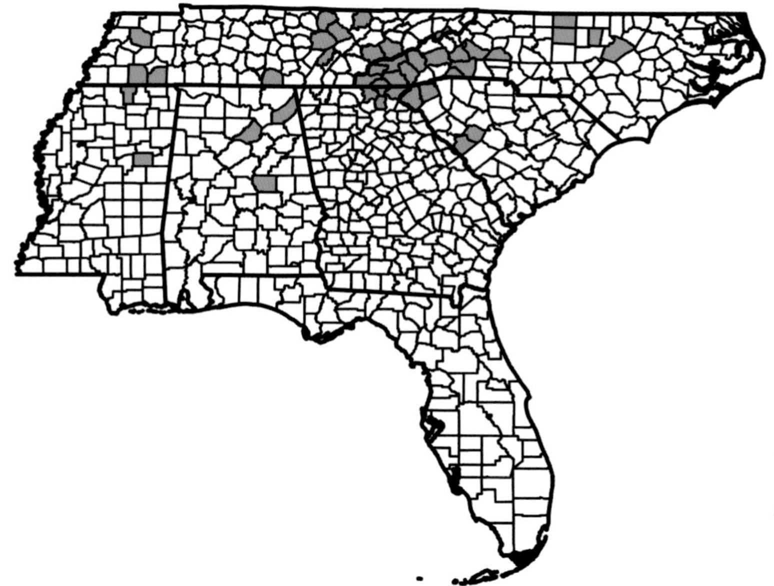

Range of the Splendid Tiger Beetle (*Cicindela splendida*) in the southeastern United States.

the posterior end. The middle band is reduced to a single short and usually straight transverse line, extending from the edge of the elytra toward the suture. Occasionally, the transverse segment is longer and has a posterior bend halfway down its length. The marginal line is absent. The apical lunule is typically reduced to a single straight line along the posterior tip of the elytra; rarely, a single anterior dot is also present. The abdomen is metallic green.

Similar Species The only species of tiger beetle in the region remotely similar is the Cow Path Tiger Beetle (*C. purpurea*), but in that species the dorsal surfaces are all the same reddish or reddish-purple color, and the transverse segment of the middle band does not reach the edge of the elytra.

Habitat Open areas of bare dirt, often with a clay component, primarily in the Blue Ridge or Piedmont. Also found along dirt service roads or forest trails, as long as they remain open and contain bare soil.

Note Finds newly open areas within a couple years of clearing through some as yet unexplained method of dispersal, and will remain at a newly colonized spot until plant succession eliminates open bare dirt. Never numerous. This species is not particularly wary and usually makes fairly short escape flights.

Oblique-lined Tiger Beetle

Cicindela tranquebarica tranquebarica Herbst

Size 12–15 mm.

Southeast Status Fairly common in the Blue Ridge and in the Coastal Plain of Mississippi and western Tennessee; may be locally common. Uncommon to rare in the Piedmont, and very rare to absent elsewhere in the Coastal Plain.

Identification A medium-sized brown tiger beetle with a large white labrum. The elytra are wide, fairly well marked, and weakly granulate. All dorsal surfaces are uniformly dark to medium brown. The humeral lunule is usually broken, with a single dot at the anterior end and a fuller mark at the posterior end; the posterior mark terminates in a distinct straight line angled toward the elytral suture. The thin middle band is typically complete but may be broken in some individuals. The marginal line is absent. The apical lunule is also thin but

Below: Flight season of the Oblique-lined Tiger Beetle (*Cicindela tranquebarica tranquebarica*).

Jan	Feb	Mar	Apr	May	Jun	Jul	Aug	Sep	Oct	Nov	Dec

Clockwise from top left:

Oblique-lined Tiger Beetle (*Cicindela tranquebarica tranquebarica*). Monroe County, Mississippi.

Oblique-lined Tiger Beetle (*Cicindela tranquebarica tranquebarica*). Talbot County, Georgia.

Oblique-lined Tiger Beetle (*Cicindela tranquebarica tranquebarica*). Towns County, Georgia.

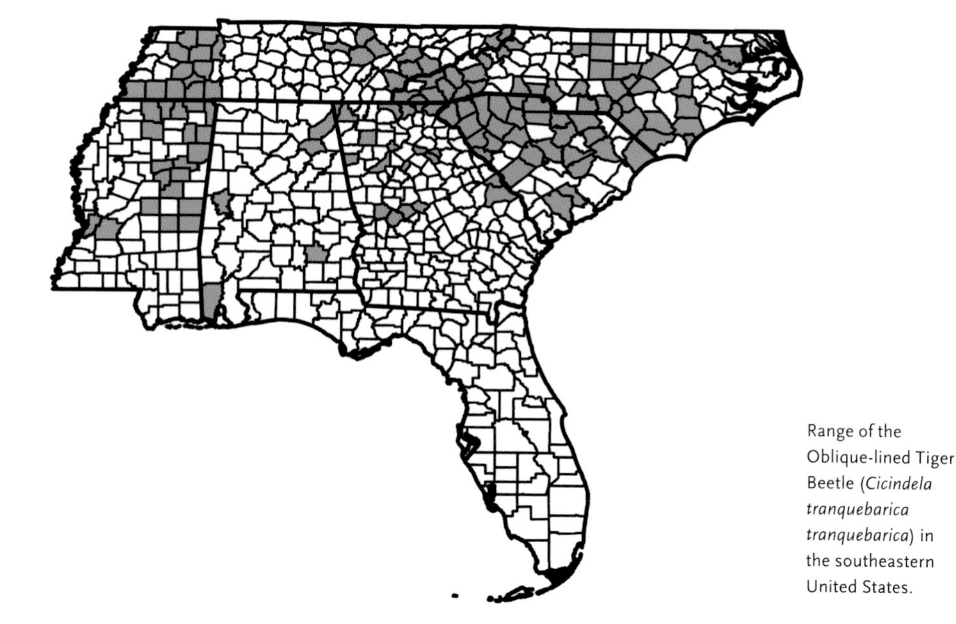

Range of the Oblique-lined Tiger Beetle (*Cicindela tranquebarica tranquebarica*) in the southeastern United States.

complete. In the fall, the markings of worn adults may be greatly reduced or even absent; however, the apical lunule is almost always still visible. The legs are brown to purple, and the abdomen varies from green to dark brown.

Similar Species The Oblique-lined Tiger Beetle is most similar to the Bronzed Tiger Beetle (*C. repanda*), Twelve-spotted Tiger Beetle (*C. duodecimguttata*), and Appalachian Tiger Beetle (*C. ancocisconensis*). All are brown or dark brown with distinct markings, but they are not usually found in the same habitat as the Oblique-lined Tiger Beetle, and they do not have the diagnostic posterior terminus of the humeral lunule.

Habitat Open sandy or bare soil areas including dirt roads and trails; not typically found near water. Tolerates very little vegetation.

Note This species very quickly colonizes new openings in the Blue Ridge, often finding sites within a year or so of clearing. Can number in the hundreds in one small patch of habitat within the first year of colonization. This is often the last species active in late fall in the Blue Ridge. The species is not very wary, but once disturbed the escape flights are long.

Eastern Pinebarrens Tiger Beetle

Cicindelidia abdominalis (Fabricius)

Size 8–11 mm.

Southeast Status Fairly common throughout the Coastal Plain of South Carolina, Georgia, Florida, and Alabama. Range also extends into southeast corners of North Carolina and Mississippi; absent from south Florida.

Identification A small, dark tiger beetle with smooth to somewhat weakly granulate, highly reflective elytra, which have reduced white markings. The labrum is large and white, typically with four long, thin setae. The dorsal color is black with few green highlights. The thorax has a narrow line of decumbent white setae along the side. The humeral lunule is absent. The middle band is typically absent or

Jan	Feb	Mar	Apr	May	Jun	Jul	Aug	Sep	Oct	Nov	Dec

Above: Flight season of the Eastern Pinebarrens Tiger Beetle (*Cicindelidia abdominalis*).

Top to bottom:

Eastern Pinebarrens Tiger Beetle (*Cicindelidia abdominalis*). Early County, Georgia.

Eastern Pinebarrens Tiger Beetle (*Cicindelidia abdominalis*). Appling County, Georgia.

reduced to two discrete spots; these spots are occasionally fused into a shortened complete mark. The marginal line is largely absent, but strongly marked individuals retain a single spot at the posterior end. The apical lunule, which is of variable thickness, is typically confined to the elytral margins, although occasional individuals possess an additional anteromedial spot. Many individuals are immaculate except for the apical lunules. Each elytron has a row of obvious punctures parallel to the suture. Scattered setae are present along the lateral portion of the anterior segments of the abdomen. The legs are metallic green or reddish brown. The reddish abdomen can be seen during flight or at rest if extended beyond the elytra.

Similar Species This tiger beetle is one of four very similar small dark *Cicindelidia* with reddish abdomens. However, with great care they can be distinguished in the field. The Eastern Pinebarrens Tiger Beetle is the northernmost species and is the most easily separated by range as it is essentially the only one found north of Florida. The Eastern Pinebarrens Tiger Beetle and Highlands Tiger Beetle (*C. highlandensis*) have smoother, less granulate elytra and typically have four labral setae. Compared with the Eastern Pinebarrens Tiger Beetle, the Highlands Tiger Beetle has a weak apical lunule and no other elytral markings, more green reflections dorsally, and no setae anywhere on the thorax or abdomen. The other two species, the Scabrous Tiger Beetle (*C. scabrosa*) and Miami Tiger Beetle (*C. floridana*), differ

Eastern Pinebarrens Tiger Beetle (*Cicindelidia abdominalis*). Lowndes County, Georgia.

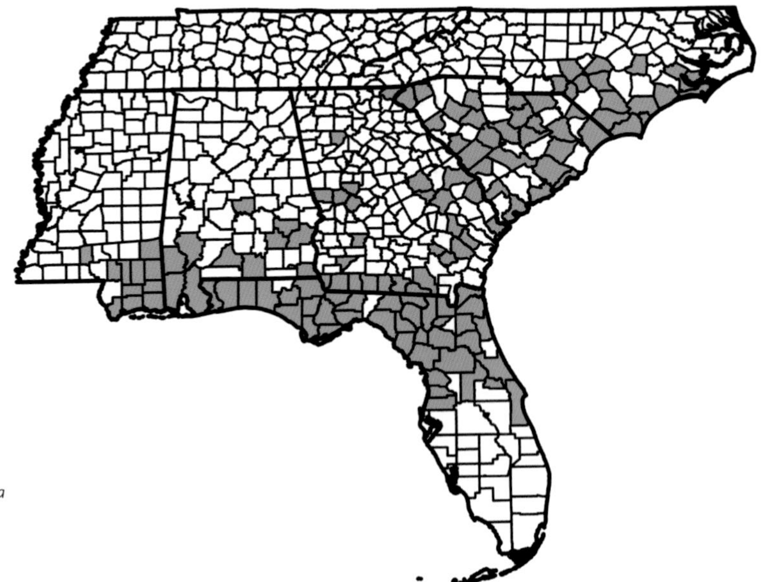

Range of the Eastern Pinebarrens Tiger Beetle (*Cicindelidia abdominalis*) in the southeastern United States.

from the Eastern Pinebarrens and Highlands Tiger Beetles by having distinctly granulate elytra and typically six labral setae. The Scabrous Tiger Beetle also has more abdominal setae than the Eastern Pinebarrens Tiger Beetle. The Miami Tiger Beetle is separated completely by range and is very greenish dorsally rather than black. The Punctured Tiger Beetle (*C. punctulata*) is also a poorly marked dark beetle, but it is larger than the Eastern Pinebarrens Tiger Beetle and does not have a reddish abdomen.

Habitat Dry sandy areas, often with scattered vegetation. Found in a variety of locations including trails through pine forests, small open areas in fields, or openings in deciduous forests.

Note This species is somewhat wary but usually makes only short escape flights. The habitats it utilizes often have small dark pieces of rocks, bark, or other vegetation in the sand, effectively providing camouflage for this small dark beetle and making it difficult to locate, even in the open.

Miami Tiger Beetle

Cicindelidia floridana (Cartwright)

Size 7–9 mm.

Southeast Status Extremely rare and local, known only from a few small sites in Miami-Dade County, Florida.

Identification A small, dark tiger beetle with distinctly granulate, largely nonreflective elytra and greatly reduced white markings. The large white labrum typically has six long, thin setae. The dorsal color is blackish with obvious widespread greenish and fewer scattered coppery highlights. The thorax has a thick line of decumbent white setae laterally. The humeral lunule, middle band, and marginal line are absent. Very rarely, a remnant of the marginal line is represented by a single posterior spot. The thin apical lunule is restricted to the elytral margins. Each elytron has a row of very obvious greenish punctures along the suture. The legs are metallic green and reddish, with some paler areas. The reddish abdomen, which is densely covered with setae along the lateral portion of the anterior segments, can be seen during flight or if extended beyond the elytra when at rest.

Similar Species This tiger beetle is one of four very similar small dark *Cicindelidia* with reddish abdomens. However, with great care they can be distinguished in the field. The Miami Tiger Beetle and Scabrous Tiger Beetle (*C. scabrosa*) have distinctly granulate elytra

Below: Flight season of the Miami Tiger Beetle (*Cicindelidia floridana*).

Jan	Feb	Mar	Apr	May	Jun	Jul	Aug	Sep	Oct	Nov	Dec

Left to right:

Miami Tiger Beetle (*Cicindelidia floridana*). Miami-Dade County, Florida.

Miami Tiger Beetle (*Cicindelidia floridana*). Miami-Dade County, Florida.

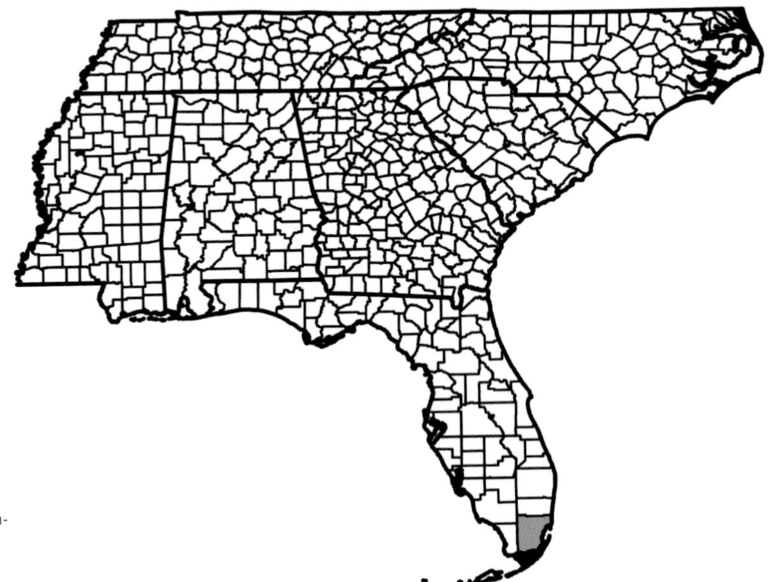

Range of the Miami Tiger Beetle (*Cicindelidia floridana*) in the southeastern United States.

and typically have six labral setae. The Scabrous Tiger Beetle is much blacker than the greenish Miami Tiger Beetle, and the Scabrous Tiger Beetle has brighter and more conspicuous elytral markings. The other two species, the Eastern Pinebarrens Tiger Beetle (*C. abdominalis*) and Highlands Tiger Beetle (*C. highlandensis*), have smoother, less granulate elytra and typically only four labral setae. The Eastern Pinebarrens Tiger Beetle has wider elytral markings and fewer lateral abdominal setae than the Miami Tiger Beetle; the Highlands Tiger Beetle is darker in color with no lateral setae on the thorax or abdomen.

Habitat Pine rocklands with saw palmetto; this beetle occupies the few small openings of limestone and sand in this habitat. Can sometimes be found in adjacent pine forest along sandy or rocky trails.

Note Not very wary, and makes short escape flights. Upon landing, immediately flattens onto substrate or up against vegetation or rocks, and can be very hard to relocate. This species is often most active during the hottest part of the day. The Miami Tiger Beetle was thought extinct, and it was only recently rediscovered and elevated to species rank (Brzoska et. al. 2011).

Highlands Tiger Beetle

Cicindelidia highlandensis (Choate)

Size 7–9 mm.

Southeast Status Rare, found only along the Lake Wales Sand Ridge in Polk and Highlands Counties, Florida.

Identification A small, dark tiger beetle with smooth to somewhat weakly granulate elytra. The labrum is large and white, and typically has four long, thin setae. The dorsal color is black with numerous green highlights. The thorax and abdomen are glabrous (without setae). Elytra are highly reflective and virtually unmarked. The humeral lunule, middle band, and marginal line are absent. The apical lunule, if present, is thin, often appears faded, and is restricted to the posterior elytral margins. Each elytron has a row of obvious punctures parallel to the suture. The legs are metallic green and red. The reddish abdomen can be seen during flight or when extended beyond the elytra at rest.

Similar Species This tiger beetle is one of four very similar small dark *Cicindelidia* with reddish abdomens. However, with great care they can be distinguished in the field. The Highlands Tiger Beetle and Eastern Pinebarrens Tiger Beetle (*C. abdominalis*) have smoother less granulate elytra and usually have four labral setae. Compared

Jan	Feb	Mar	Apr	May	Jun	Jul	Aug	Sep	Oct	Nov	Dec

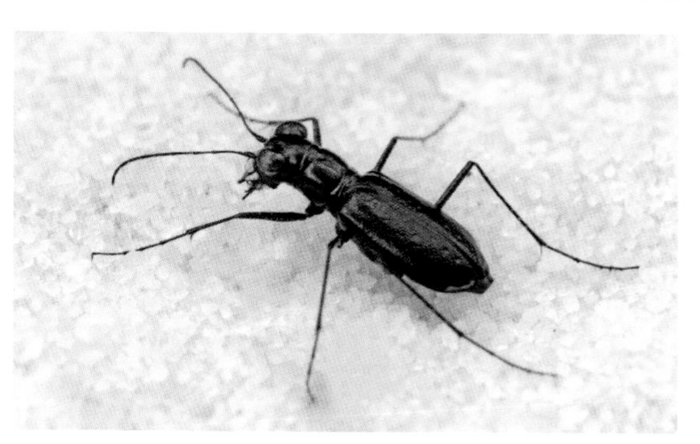

Above: Flight season of the Highlands Tiger Beetle (*Cicindelidia highlandensis*).

Left: Highlands Tiger Beetle (*Cicindelidia highlandensis*). Highlands County, Florida.

Top to bottom:

Highlands Tiger Beetle (*Cicindelidia highlandensis*). Polk County, Florida.

Highlands Tiger Beetle (*Cicindelidia highlandensis*). Highlands County, Florida.

with the Eastern Pinebarrens Tiger Beetle, the Highlands Tiger Beetle has a weak apical lunule and no other elytral markings, more green reflections dorsally, and no setae anywhere on the thorax or abdomen. The other two species, Scabrous Tiger Beetle (*C. scabrosa*) and Miami Tiger Beetle (*C. floridana*), differ from the Highlands and Eastern Pinebarrens Tiger Beetles by having distinctly granulate elytra and usually six labral setae. The Scabrous Tiger Beetle also has a much more distinct apical lunule and more thoracic and abdominal setae than Highlands Tiger Beetle. The Miami Tiger Beetle is separated completely by range and is very greenish dorsally rather than black. The Punctured Tiger Beetle (*C. punctulata*) is also a poorly marked dark beetle, but it is larger than the Highlands Tiger Beetle and does not have a reddish abdomen.

Habitat Deep loose white sand openings within or adjacent to pine scrub, pine flatwoods or longleaf pine forest. Habitat loss due to development and expansion of citrus groves has greatly restricted available habitat for this species, although it is now protected at several key sites (Knisley 2013).

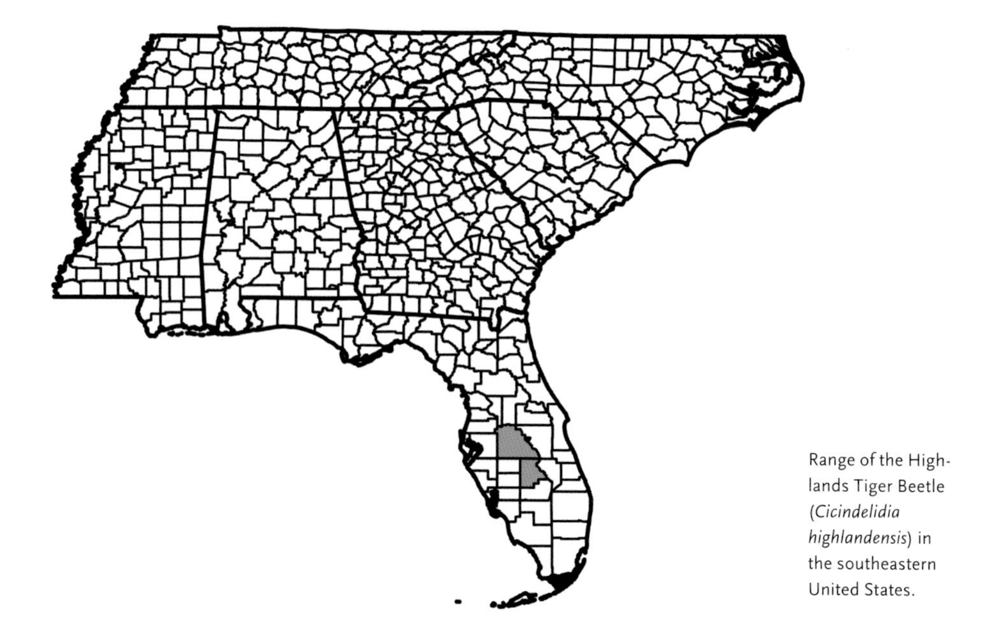

Range of the Highlands Tiger Beetle (*Cicindelidia highlandensis*) in the southeastern United States.

NOTE The Highlands Tiger Beetle is not wary and makes short escape flights when disturbed. This species seeks shade in vegetation during the hottest part of the day. It is relatively new to science, having been discovered and formally described in 1984 by the late Paul "Skip" Choate (Choate 1984).

Cobblestone Tiger Beetle

Cicindelidia marginipennis (Dejean)

Size 11–14 mm.

Southeast Status Extremely rare and local; known from very few sites in the region.

Identification A medium-sized brown tiger beetle with unique elytral markings. The head and pronotum are dark brown, with some red or green reflective areas, and the thorax has a lateral row of setae. The brown elytra have an expanded white or cream marginal line along their entire length, from the humeral angle to the elytral tip. This line encompasses the humeral and apical lunules and the middle band, which lacks any transverse markings. There are small medial projections from the marginal line at the middle band and apical lunule. The legs are metallic green and coppery red. The reddish-orange abdomen can easily be seen during flight or if extended beyond the elytra when at rest.

Below: Flight season of the Cobblestone Tiger Beetle (*Cicindelidia marginipennis*).

Jan	Feb	Mar	Apr	May	Jun	Jul	Aug	Sep	Oct	Nov	Dec

Clockwise from top left:

Cobblestone Tiger Beetle (*Cicindelidia marginipennis*). Perry County, Alabama.

Cobblestone Tiger Beetle (*Cicindelidia marginipennis*). Perry County, Alabama.

Cobblestone Tiger Beetle (*Cicindelidia marginipennis*). Perry County, Alabama.

Northern form Cobblestone Tiger Beetle (*Cicindelidia marginipennis*). Southern Kentucky.

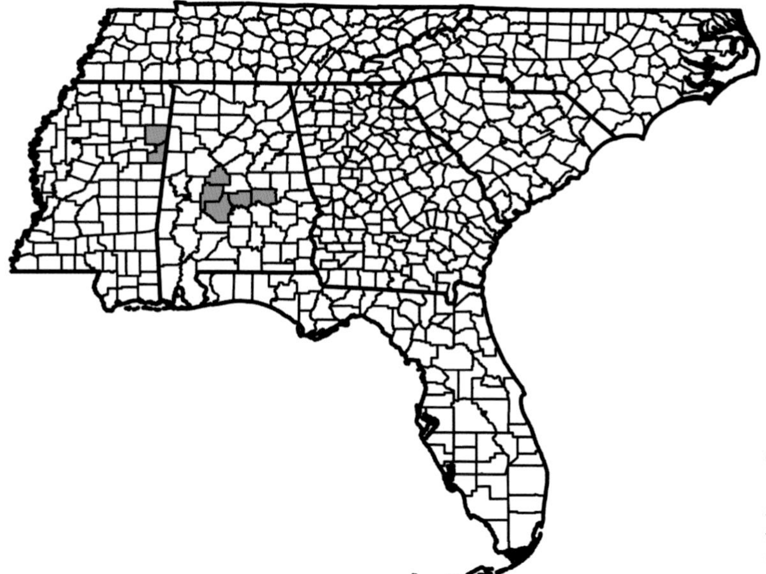

Range of the Cobblestone Tiger Beetle (*Cicindelidia marginipennis*) in the southeastern United States.

Similar Species No other tiger beetle in the Southeast resembles this species.

Habitat Typically restricted to cobblestone river bars, but may be found in adjacent open habitat during periods of high water. The rarity of this habitat undoubtedly limits the distribution of this species in the region.

Note This is among the wariest of tiger beetles, and it almost always makes long and unpredictable escape flights. For these reasons, plus its rarity, the Cobblestone Tiger Beetle is very difficult to observe or photograph. It moves frequently and rapidly when on the ground and often climbs up on small cobblestone rocks for better visibility. It may be locally numerous in high quality habitats. This species occurs in two disjunct areas in the US: roughly from Kentucky and Indiana east and then north into New England, and in Alabama and Mississippi in the Southeast. Individuals in the Southeast tend to be darker brown and larger, with a wider marginal line, than those farther to the north. Recent molecular work suggests that these may be two different species.

Ocellated Tiger Beetle

Cicindelidia ocellata rectilatera (Chaudoir)

Size 9–13 mm.

Southeast Status Very rare in the region; known only from one site along the banks of the Mississippi River.

Identification A medium-sized dark brown tiger beetle with elytral maculations reduced to eight round white spots. The head and pronotum have some small areas with green reflections. The weakly granulate elytra are mostly nonreflective. The humeral lunule is reduced to a single large spot at its posterior end. The middle band is reduced to one large spot at each end of the transverse segment. The marginal line is absent. The apical lunule is reduced to a single large spot at its anterior end. The legs are metallic green and brown, and the abdomen is green to dark reddish.

Below: Flight season of the Ocellated Tiger Beetle (*Cicindelidia ocellata rectilatera*).

Jan	Feb	Mar	Apr	May	Jun	Jul	Aug	Sep	Oct	Nov	Dec

Clockwise from top left:

Ocellated Tiger Beetle (*Cicindelidia ocellata rectilatera*). Issaquena County, Mississippi.

Ocellated Tiger Beetle (*Cicindelidia ocellata rectilatera*). Issaquena County, Mississippi.

Ocellated Tiger Beetle (*Cicindelidia ocellata rectilatera*). Travis County, Texas.

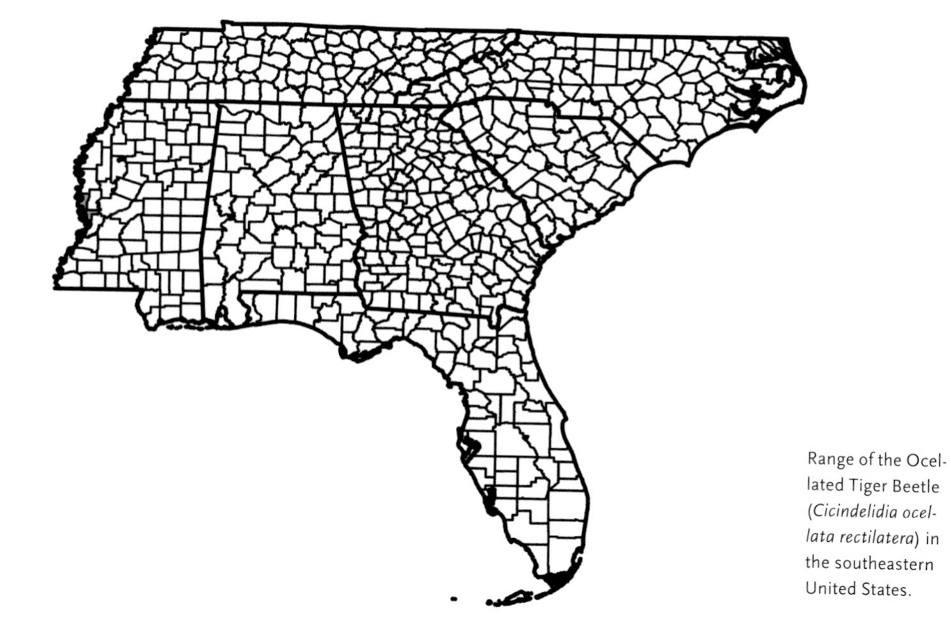

Range of the Ocellated Tiger Beetle (*Cicindelidia ocellata rectilatera*) in the southeastern United States.

Similar Species All other brown tiger beetles in the region have either smaller spots or more elongated maculations with more connecting lines.

Habitat Sandy or muddy bars along large rivers. In the rest of its range it may also be found along wet pond edges. Observed along sandy seams in broken asphalt at the one known site in the region.

Note This species is not very wary and makes fairly short escape flights. The Ocellated Tiger Beetle is much more common in the southwestern United States, where it may be very numerous at sites with its preferred habitat.

Punctured Tiger Beetle

Cicindelidia punctulata punctulata (Olivier)

Size 10–13 mm.

Southeast Status Very common throughout the region, but not numerous at any single location. Found in even the most altered habitat, as long as it is open.

Identification A medium-sized but slender dark brown tiger beetle with limited elytral markings. The head and pronotum are brown to dark brown with some mostly green metallic reflections; the labrum is white. The thorax often has a lateral line of setae. The elytra are also brown, somewhat reflective, and moderately granulate; an obvious row of greenish punctures parallels the suture on each side. White markings are always incomplete and are often largely or completely absent. The humeral lunule is typically absent; if present, it consists of only the posterior dot. The middle band, if present, is

Below: Flight season of the Punctured Tiger Beetle (*Cicindelidia punctulata punctulata*).

Jan	Feb	Mar	Apr	May	Jun	Jul	Aug	Sep	Oct	Nov	Dec

Clockwise from top left:

Punctured Tiger Beetle (*Cicindelidia punctulata punctulata*). Bolivar County, Mississippi.

Punctured Tiger Beetle (*Cicindelidia punctulata punctulata*). Lowndes County, Georgia.

Punctured Tiger Beetle (*Cicindelidia punctulata punctulata*). Laurens County, Georgia.

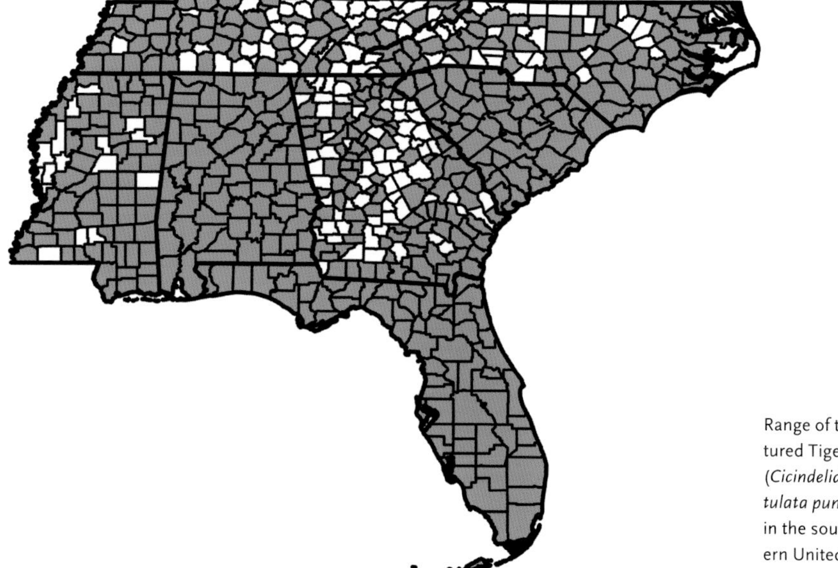

Range of the Punctured Tiger Beetle (*Cicindelidia punctulata punctulata*) in the southeastern United States.

always broken and usually consists of a series of very small spots with no connecting lines. The marginal line is either completely absent or represented by a posterior spot. The apical lunule is usually present and complete, but very thin. The legs are dark brown with few green or reddish highlights, and the abdomen is dark metallic blue or green to black.

Similar Species The Punctured Tiger Beetle can be confused with other brown and typically marked tiger beetles, but it is more slender, less strongly marked, and found in different habitats than most. The Punctured Tiger Beetle resembles the Eastern Red-bellied Tiger Beetle (*C. rufiventris* sspp.), but that species has a nonreflective "orange peel" elytral surface and a reddish abdomen. Weakly marked Punctured Tiger Beetles are similar to weakly marked S-banded Tiger Beetles (*C. trifasciata*) but are more slender with a more obvious row of elytral punctures. Additionally, the Punctured Tiger Beetle is nearly always found in dry habitat, where the S-banded Tiger Beetle is nearly always found in moist habitat.

Habitat Very dry open sandy areas, with little or no vegetation. This may include clearings, natural openings, roads and roadsides, and trails. Can inhabit dry sandbar habitats along streams and rivers, and the adults may occasionally wander fairly close to the water's edge. Manmade disturbance does not deter this species, which is sometimes referred to as the "Sidewalk Tiger Beetle."

Note The habitat favored by the Punctured Tiger Beetle is often very hot during the day; this species escapes the heat by hiding in shade or climbing on vegetation just above ground level. Usually found scattered in small numbers within suitable habitat. Not very wary, the Punctured Tiger Beetle will allow fairly close approach if the observer moves slowly, but when disturbed, it typically makes short escape flights. This species is often attracted to artificial lights at night, such as those at shopping malls and gas stations.

Eastern Red-bellied Tiger Beetle

Cicindelidia rufiventris rufiventris (Dejean)

Size 9–12 mm.

Southeast Status Occurs throughout most of the region but is absent from virtually all of Florida and much of south Georgia. Can be fairly common in some areas and uncommon in others; never abundant at any one site.

Identification A small to medium-sized brown tiger beetle with unusual and nonreflective "orange peel" elytra. The head and pronotum sometimes have small areas of coppery reflections, and the sides of the thorax are edged with setae. The labrum is white. White elytral markings are typically broken and often absent except for the apical lunule. The humeral lunule, if present, is reduced to a single anterior or posterior spot; both spots are rarely present. The middle band usually consists of one or two spots or is absent. The marginal line is typically absent, but there is sometimes a spot representing the posterior

Below: Flight season of the Eastern Red-bellied Tiger Beetle (*Cicindelidia rufiventris rufiventris*).

Jan	Feb	Mar	Apr	May	Jun	Jul	Aug	Sep	Oct	Nov	Dec

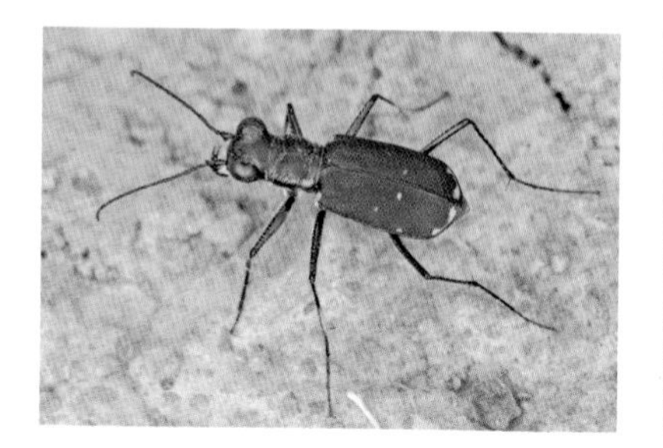

Clockwise from top left:

Eastern Red-bellied Tiger Beetle (*Cicindelidia rufiventris rufiventris*). Hinds County, Mississippi.

Eastern Red-bellied Tiger Beetle (*Cicindelidia rufiventris rufiventris*). Hinds County, Mississippi.

Pair of Eastern Red-bellied Tiger Beetles (*Cicindelidia rufiventris rufiventris*). Hinds County, Mississippi.

Left to right:

Eastern Red-bellied Tiger Beetle (*Cicindelidia rufiventris rufiventris*). Wilson County, Tennessee.

Eastern Red-bellied Tiger Beetle (*Cicindelidia rufiventris cumatilis*). Rabun County, Georgia.

terminus of this mark. The apical lunule is almost always present but thin, and it is usually reduced to the portion along the elytral margin. Infrequently there is an additional anterior apical lunule spot separated from the margin. The legs are metallic green, and the abdomen is reddish. The abdomen can often be seen during flight or extending past the tips of the elytra while the beetle is at rest.

There is another named subspecies of Eastern Red-bellied Tiger Beetle: *C. r. cumatilis* (LeConte). This subspecies is thought to occur in the western part of this beetle's range—into western Mississippi and Tennessee. However, the primary feature used to distinguish this subspecies from the nominate Eastern Red-bellied Tiger Beetle is that the elytra are bluish, and blue-backed examples are known to occur as far east as northern Georgia. It is questionable whether *C. r. cumatilis* is a valid subspecies.

Similar Species This often weakly marked species can be confused with the Punctured Tiger Beetle (*C. punctulata*), which is also weakly marked, but the Punctured Tiger Beetle has more reflective elytra with obvious rows of punctures, is larger, and has a dark abdomen. The Eastern Pinebarrens Tiger Beetle (*C. abdominalis*) overlaps somewhat in range and also has a reddish abdomen, but it is smaller and darker with much more reflective elytra. The Twelve-spotted Tiger Beetle (*Cicindela duodecimguttata*) has similar broken markings, but it is larger and does not have a reddish abdomen.

Habitat Open bare dirt with or without rocks or gravel; tolerates very little vegetation. Commonly found on clay substrates in the Southeast.

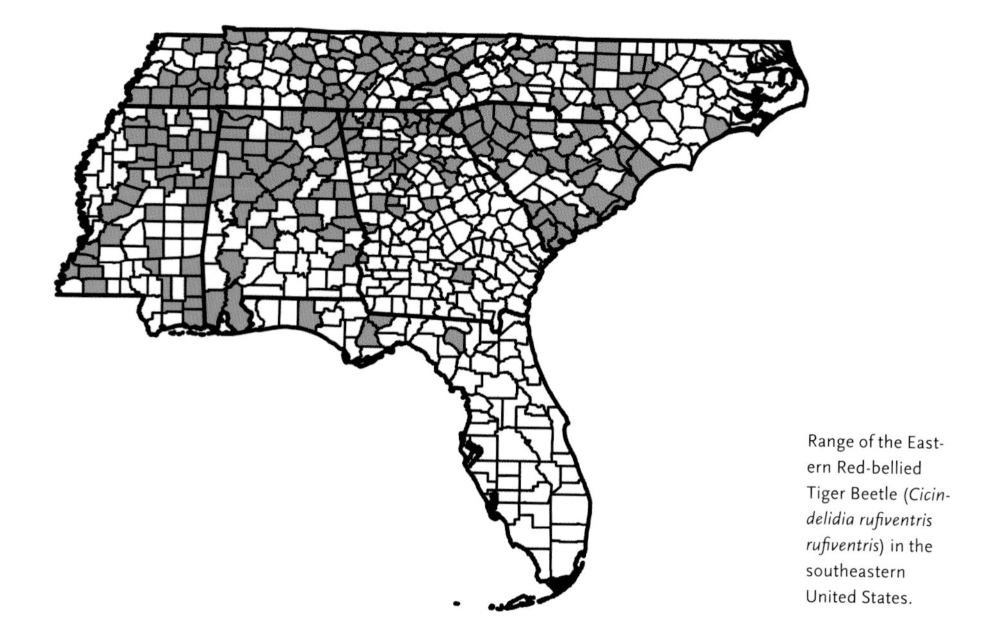

Range of the Eastern Red-bellied Tiger Beetle (*Cicindelidia rufiventris rufiventris*) in the southeastern United States.

Note The Eastern Red-bellied Tiger Beetle is not very wary and makes short, slow escape flights. It often flattens down against the ground and remains motionless upon landing. While individuals will occasionally climb up on small rocks for better visibility, this species generally moves around less on the ground than most other tiger beetles. These traits, combined with their small size, make Eastern Red-bellied Tiger Beetles surprisingly difficult to observe.

Scabrous Tiger Beetle

Cicindelidia scabrosa (Schaupp)

Size 7–8 mm.

Southeast Status Uncommon species endemic to peninsular Florida; populations highly localized where appropriate habitat exists. Never numerous.

Identification A small, dark tiger beetle with distinctly granulate elytra; the dorsal color is black with few green highlights. The labrum is large and white, usually with six long, thin setae. The thorax has a thick line of decumbent white setae laterally. The elytra are weakly reflective and marked. The humeral lunule is absent. The middle band is typically absent; if present, it is reduced to two discrete spots, which are rarely connected by a thin line. The marginal line is reduced to a single wide posterior spot on most specimens. The apical lunule is typically thick and confined to the elytral margins; an additional

Below: Flight season of the Scabrous Tiger Beetle (*Cicindelidia scabrosa*).

Jan	Feb	Mar	Apr	May	Jun	Jul	Aug	Sep	Oct	Nov	Dec

Clockwise from top left:

Scabrous Tiger Beetle (*Cicindelidia scabrosa*). Levy County, Florida.

Scabrous Tiger Beetle (*Cicindelidia scabrosa*). Levy County, Florida.

Scabrous Tiger Beetle (*Cicindelidia scabrosa*). Dixie County, Florida.

Scabrous Tiger Beetle (*Cicindelidia scabrosa*). Hardee County, Florida.

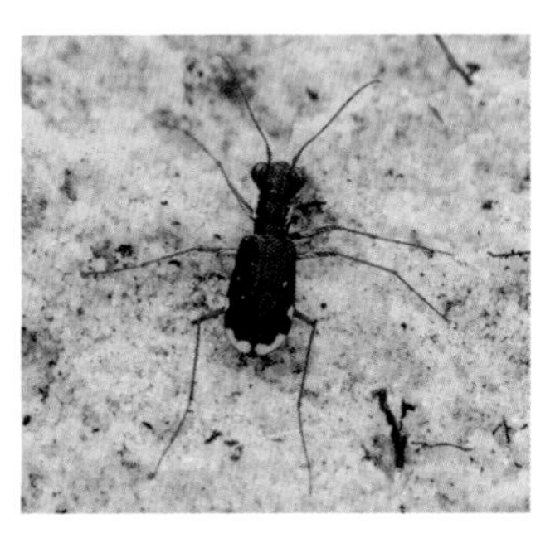

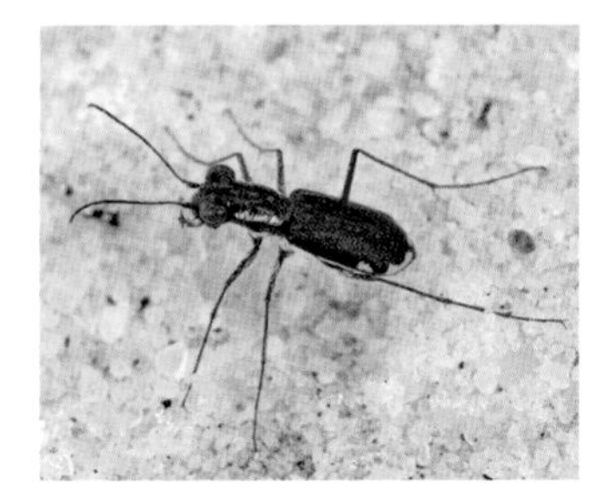

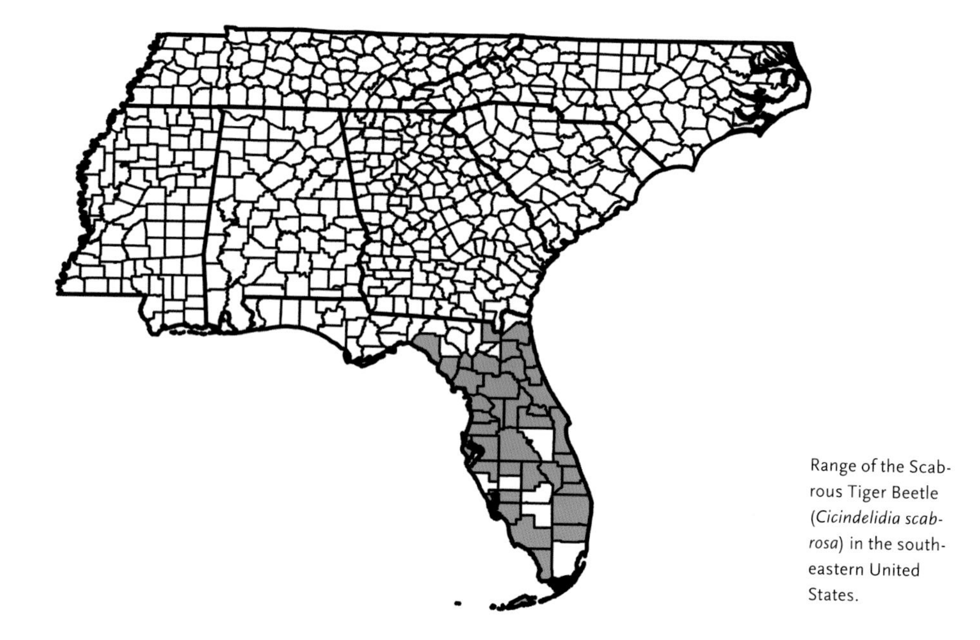

Range of the Scabrous Tiger Beetle (*Cicindelidia scabrosa*) in the southeastern United States.

anterior medial spot is sometimes present. Many individuals are immaculate except for the remnant marginal line spots and the apical lunules. Each elytron has a row of obvious punctures parallel to the suture. The legs are metallic green and reddish. The anterior abdominal segments are densely covered with white setae laterally. The reddish abdomen can be seen during flight or at rest if extended beyond the elytra.

Similar Species This tiger beetle is one of four very similar small dark *Cicindelidia* with reddish abdomens. However, with great care they can be distinguished in the field. The Scabrous Tiger Beetle and Miami Tiger Beetle (*C. floridana*) have distinctly granulate elytra and typically have six labral setae. The Scabrous Tiger Beetle is much blacker than the greenish Miami Tiger Beetle, and the Miami Tiger Beetle usually lacks any middle band or marginal line markings on the elytra. The other two species, Eastern Pinebarrens Tiger Beetle (*C. abdominalis*) and Highlands Tiger Beetle (*C. highlandensis*), have smoother elytra and usually have four labral setae. In addition to those two features, compared with the Scabrous Tiger Beetle, the Eastern Pinebarrens Tiger Beetle has fewer abdominal setae and the Highlands Tiger Beetle has fewer elytral markings and no abdominal setae. The Punctured Tiger Beetle (*C. punctulata*) is also poorly marked, but it is larger and does not have a reddish abdomen.

Habitat Loose white sand with small oaks, sometimes near pine flatwoods or other wet lowlands. Often in sand pine scrub openings and along sandy trails.

Note This small tiger beetle is not wary; when disturbed, it makes short escape flights and often flattens against the ground upon landing. It can be found during hotter parts of the day, often in vegetation just above the sand.

S-banded Tiger Beetle

Cicindelidia trifasciata ascendens (LeConte)

Size 11–13 mm.

Southeast Status Common to abundant along all coastlines throughout the region. Increasingly found far inland, mostly along river corridors.

Identification A medium-sized brown tiger beetle with unique elytral markings. All dorsal surfaces are somewhat metallic and brown, ranging in color from medium to very dark brown. The head and pronotum have limited green reflections, if any. The labrum is white to light brown. The thorax usually has a line of setae laterally. The elytra are somewhat granulate, with a row of relatively inconspicuous punctures paralleling the suture. Markings vary from wide and complete

Below: Flight season of the S-banded Tiger Beetle (*Cicindelidia trifasciata ascendens*).

Clockwise from top left:

S-banded Tiger Beetle (*Cicindelidia trifasciata ascendens*). Marengo County, Alabama.

S-banded Tiger Beetle (*Cicindelidia trifasciata ascendens*). Marengo County, Alabama.

S-banded Tiger Beetle (*Cicindelidia trifasciata ascendens*). Appling County, Georgia.

S-banded Tiger Beetle (*Cicindelidia trifasciata ascendens*). Appling County, Georgia.

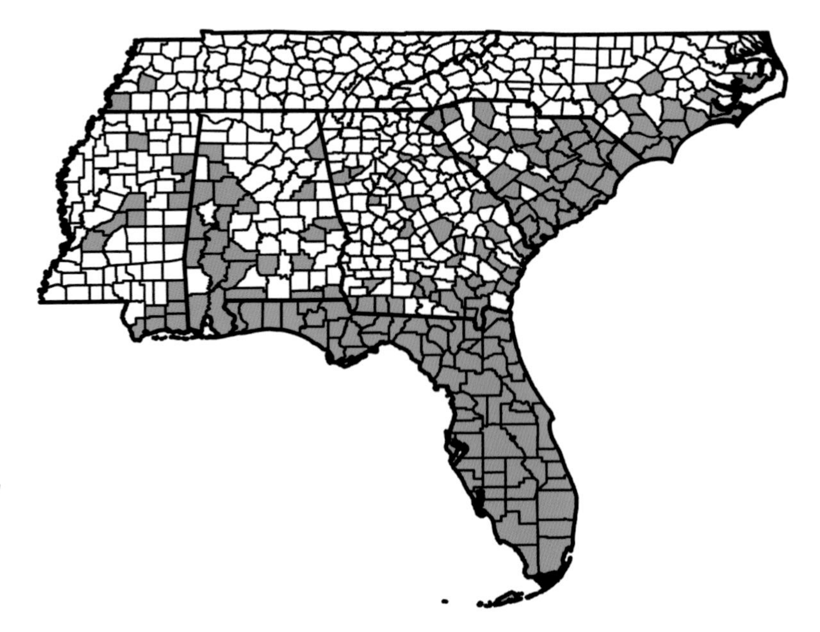

Range of the S-banded Tiger Beetle (*Cicindelidia trifasciata ascendens*) in the southeastern United States.

to almost completely lacking but are most often thin and partially broken. The color of markings varies from pure white to cream or light brown. A complete humeral lunule consists of a thin sweeping curve ending in a straight transverse line with a small, expanded spot at the posteromedial end; in most individuals, however, only the posterior spot is present. The middle band is long and traces a very obvious and rounded S shape but is often faded, incomplete, or otherwise difficult to see. When present, the shape of this mark is diagnostic for this species. There is occasionally a marginal line extending rearward from the middle band and ending in an expanded spot, but more often only the spot is present. The thin apical lunule is typically complete. The legs are brown or black and often have reddish or greenish metallic highlights. The abdomen is blackish blue.

Similar Species Well-marked individuals are easily identified by the S-shaped middle band, but poorly marked individuals can at times be confused with other lightly marked brown tiger beetles. Most often confused with the Punctured Tiger Beetle (*C. punctulata*) but is not as slender as that species; the S-banded Tiger Beetle is nearly always found on moist or muddy substrate, while the Punctured Tiger Beetle is almost always found on dry substrate. Rows of elytral punctures on the S-banded Tiger Beetle are not as obvious as those on the

Punctured Tiger Beetle. Can resemble Eastern Red-bellied Tiger Beetle (*C. rufiventris* sspp.), but that species has a nonreflective "orange peel" elytral surface and a reddish abdomen, and is usually found on drier soils in upland habitats.

Habitat Almost exclusively found on moist or muddy surfaces, ranging from wet sand and marsh mud along the coast to muddy edges of inland rivers and ponds.

Note This species may be abundant at coastal sites but is typically found in low numbers inland. Almost all historical records are from along the coast; more recently, the S-banded Tiger Beetle is increasingly being found inland throughout the region, and it is by far the most likely species of tiger beetle to be found outside its mapped range. This species is clearly capable of long-range dispersal flights, and individuals have been recorded more than 320 km inland from its normal range. It has also been recorded 160 km offshore on an oil platform in the Gulf of Mexico (Graves 1981). The S-banded Tiger Beetle is a very wary species, but it generally makes rather short escape flights.

Sandbar Tiger Beetle

Ellipsoptera blanda (Dejean)

Size 11–13 mm.

Southeast Status Uncommon to fairly common along sandy streams and rivers in lower coastal plains of the region from southeastern Mississippi to southeastern North Carolina.

Identification A medium-sized light brown or bronze tiger beetle with extensive elytral markings. The dorsal surface of the head and pronotum are sparsely to densely covered with decumbent white setae. The sides of the head, thorax, and abdomen are densely setose. The labrum is usually ivory. The elytra are weakly granulate, with extensive white or cream-colored markings. The humeral lunule is

Jan	Feb	Mar	Apr	May	Jun	Jul	Aug	Sep	Oct	Nov	Dec

Above: Flight season of the Sandbar Tiger Beetle (*Ellipsoptera blanda*).

Clockwise from top left:

Pair of Sandbar Tiger Beetles (*Ellipsoptera blanda*). Santa Rosa County, Florida.

Pair of Sandbar Tiger Beetles (*Ellipsoptera blanda*). Long County, Georgia.

Sandbar Tiger Beetle (*Ellipsoptera blanda*). Long County, Georgia.

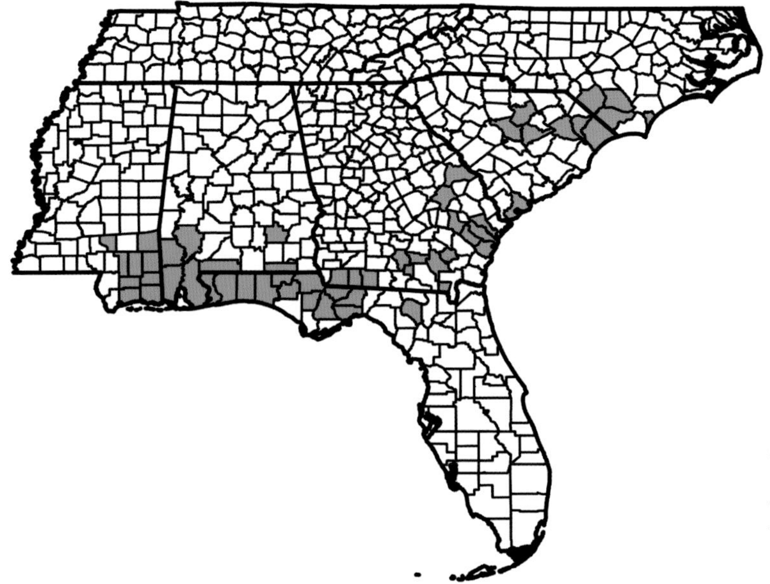

Range of the Sandbar Tiger Beetle (*Ellipsoptera blanda*) in the southeastern United States.

usually complete and strongly curved medially with expanded spots at both ends, resembling a capital G. The middle band is long and expanded medially, as is typical for this genus. From the marginal line, it curves forward toward the posterior portion of the humeral lunule before turning sharply rearward and expanding greatly into an area of diffuse wavy lines somewhat resembling a kidney or pea pod in shape. The apical lunule is complete and also extends anteriorly along the elytral suture. The marginal line is thick and is complete except for the area between the anterior portion of both humeral lunules. All elytral markings are thickened and wide, covering much of the elytra. The legs are metallic greenish or reddish brown, and the femora are sometimes covered with white setae. The abdomen is usually bright metallic green.

Similar Species The Sandbar Tiger Beetle is somewhat similar to several other members of the genus *Ellipsoptera* but is usually the only river species found in the eastern part of the region. Its range abuts that of the Coppery Tiger Beetle (*E. cuprascens*) in Alabama and Mississippi, but the Sandbar Tiger Beetle is much more extensively marked than the Coppery Tiger Beetle and has a paler background elytra color. The Sandy Stream Tiger Beetle (*E. macra*) is also somewhat similar but the ranges do not overlap. The Ghost Tiger Beetle (*E. lepida*) is fairly similar but has even more expanded elytral markings and pale legs; it is almost always found in drier habitats well away from the water's edge. The ranges of the Sandbar and Ghost Tiger Beetles only overlap in a small area of southeast Mississippi. The Coastal Tiger Beetle (*E. hamata* sspp.) and Margined Tiger Beetle (*E. marginata*) are also somewhat similar, but they are restricted in habitat to coastal shores where the Sandbar Tiger Beetle is not found. The White Sand Tiger Beetle (*E. wapleri*) is often found in the same streams in the western portion of the range, but it is a smaller and darker species with much less extensive elytral markings.

Habitat Large streams and rivers with sandy substrates; occasionally on other substrates in larger rivers.

Note This species is very wary and is one of the more difficult tiger beetles to photograph or net. It is most often found right along the water's edge, but in cooler parts of the day it may also be found away from water on drier parts of sandbars. Heavy rainfall and subsequent flooding may cause this species to move into alternate habitats, such as sand roads and other openings near the flooded stream.

Coppery Tiger Beetle

Ellipsoptera cuprascens (LeConte)

Size 10–14 mm.

Southeast Status Fairly common along large streams or rivers in Mississippi and western Alabama; known in Tennessee only from far western counties along Mississippi River.

Identification A medium-sized brown to light brown tiger beetle with distinct elytral markings. The dorsal surface of the head and pronotum are sparsely covered with decumbent white setae. The sides of the head, thorax, and abdomen are densely setose. The labrum is ivory to beige. The elytra are usually strongly granulate and typically have well-defined white or cream-colored markings; the apices are pointed in both sexes. The humeral lunule is usually complete and strongly curved medially with expanded spots at both ends, resembling a

Below: Flight season of the Coppery Tiger Beetle (*Ellipsoptera cuprascens*).

Clockwise from top left:

Coppery Tiger Beetle (*Ellipsoptera cuprascens*). Greene County, Alabama.

Coppery Tiger Beetle (*Ellipsoptera cuprascens*). Greene County, Alabama.

Coppery Tiger Beetle (*Ellipsoptera cuprascens*). Sumter County, Alabama.

Coppery Tiger Beetle (*Ellipsoptera cuprascens*). Bolivar County, Mississippi.

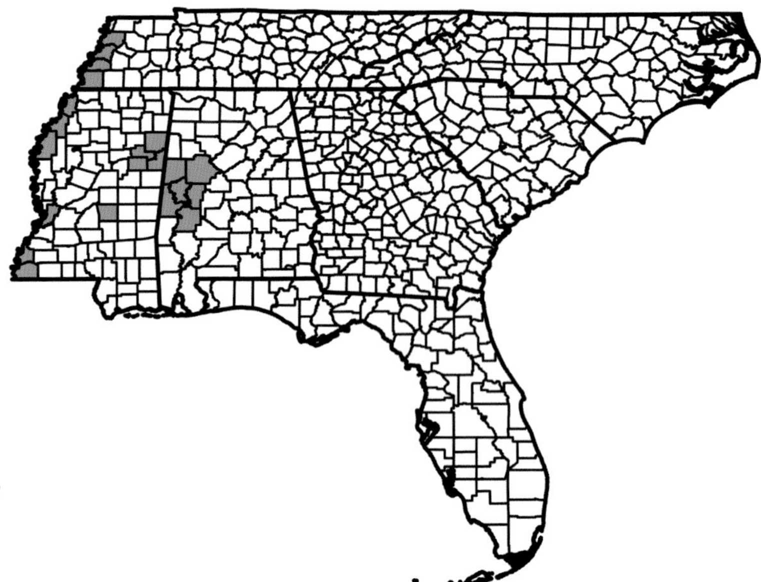

Range of the Coppery Tiger Beetle (*Ellipsoptera cuprascens*) in the southeastern United States.

capital G. The middle band is long and expanded medially, as is typical for this genus. From the marginal line, it curves forward toward the posterior portion of the humeral lunule before turning sharply rearward and expanding slightly at the distal tip. Portions of this marking, especially between the rearward turn and distal tip, may be incomplete. The apical lunule is typically thick and complete, and also extends anteriorly along the elytral suture. The marginal line varies in thickness and is almost always complete except for the area between the anterior portions of both humeral lunules. The legs are metallic greenish or reddish brown, and the femora are sometimes covered with white setae. The abdomen is reddish brown to green.

Similar Species The Sandy Stream Tiger Beetle (*E. macra*) is extremely similar but has weakly granulate elytra. Additionally, females can be separated by carefully examining the elytra: the apices, which are visible in the field with close-focusing binoculars or through a high-quality camera lens, are sharply pointed in Coppery Tiger Beetles and rounded in Sandy Stream Tiger Beetles. Some sources cite the color of the elytra as a distinguishing feature; this character is more helpful when separating more western populations of Coppery Tiger Beetles that are, as the common name suggests, more obviously coppery in color. The more eastern form occurring in the Southeast is less coppery and browner, like the Sandy Stream Tiger Beetle. Additionally, the Coppery Tiger Beetle is far more likely to be encountered in the region than the Sandy Stream Tiger Beetle, even along the Mississippi River. Other members of this genus, such as the Coastal Tiger Beetle (*E. hamata* sspp.), Margined Tiger Beetle (*E. marginata*), and Sandbar Tiger Beetle (*E. blanda*), are also very similar. However, the Coastal and Margined Tiger Beetles are restricted to coastal shores, where the Coppery Tiger Beetle is not found. The range of the Sandbar Tiger Beetle is mostly farther east, but the ranges of the two species meet in south Mississippi and Alabama. The Sandbar Tiger Beetle typically has much thicker elytral markings than the Coppery Tiger Beetle, and the abdomen is a bright metallic green. Due to differences in preferred habitat, it is very unlikely that these two species would occur together.

Habitat Fairly large streams and small to large rivers with sandy substrates.

Note The Coppery Tiger Beetle is so similar to the Sandy Stream Tiger Beetle that even museum records from the Southeast are often mislabeled; literature records may not always be reliable. The eastern form of the Coppery Tiger Beetle was formerly recognized as the subspecies *E. c. mundula* (Casey) and may appear as such in some of the older literature.

Whitish Tiger Beetle

Ellipsoptera gratiosa (Guérin-Méneville)

Size 10–12 mm.

Southeast Status Uncommon, with populations occupying two disjunct areas within the Coastal Plain: the Carolinas, and extreme southern Alabama and the western panhandle of Florida. There is one historical record for Georgia based on a specimen collected in 1941 in the extreme southwestern part of the state.

Identification A small to medium-sized pale white tiger beetle. The dorsal surface of the head and pronotum are densely covered with decumbent white setae. The sides of the head, thorax, and abdomen are also densely setose. Setae are absent from the dorsolateral edges of the thorax, giving the appearance of two dark lines due to their underlying dark color. The labrum is ivory with very few or no decumbent setae. White maculations are greatly expanded, leaving only a dark

Below: Flight season of the Whitish Tiger Beetle (*Ellipsoptera gratiosa*).

Jan	Feb	Mar	Apr	May	Jun	Jul	Aug	Sep	Oct	Nov	Dec

Clockwise from top left:

Whitish Tiger Beetle (*Ellipsoptera gratiosa*). Bay County, Florida.

Whitish Tiger Beetle (*Ellipsoptera gratiosa*). Washington County, Florida.

Close-up of the labrum of a Whitish Tiger Beetle (*Ellipsoptera gratiosa*). Lexington County, South Carolina.

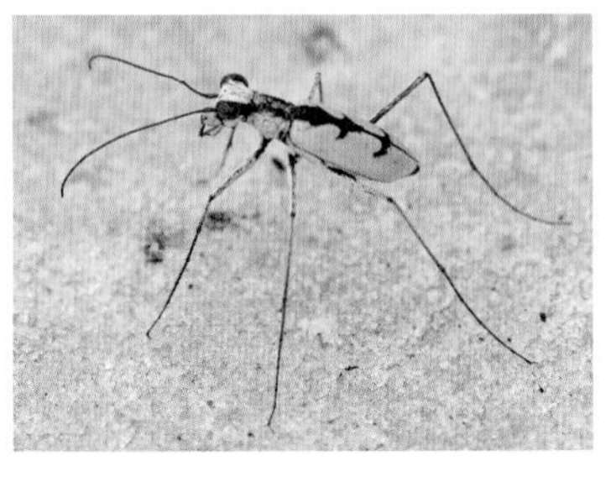

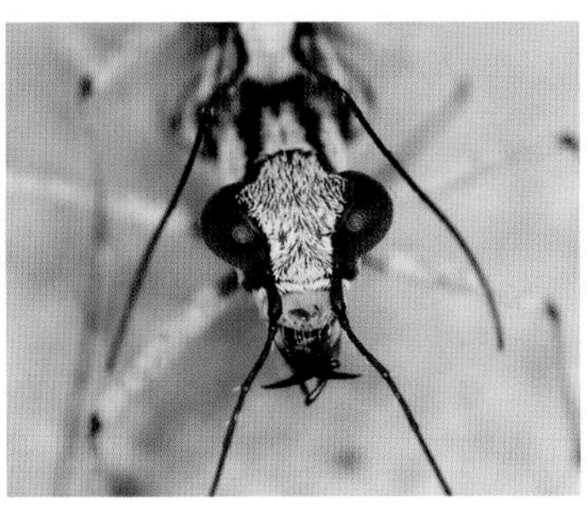

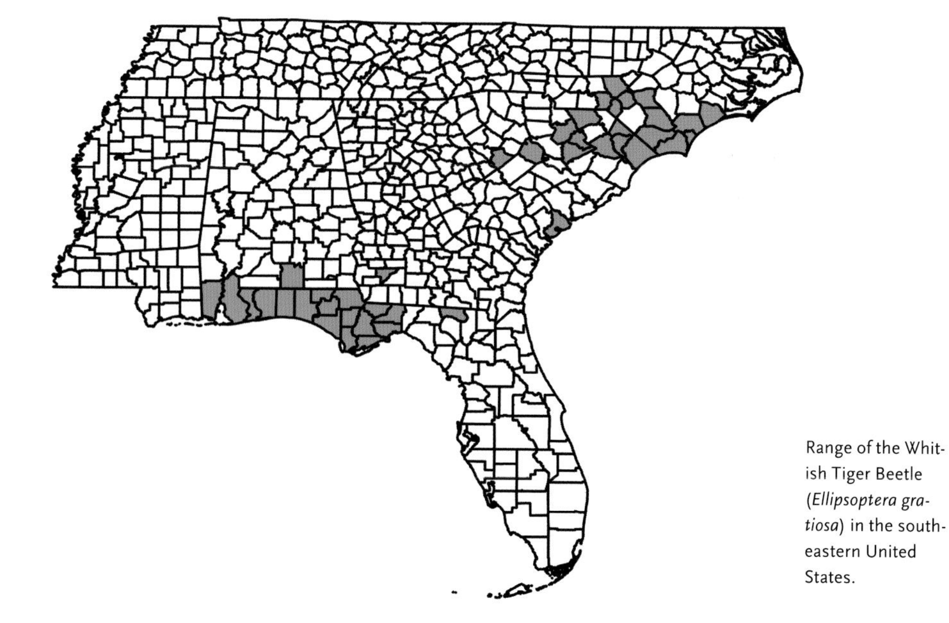

Range of the Whitish Tiger Beetle (*Ellipsoptera gratiosa*) in the southeastern United States.

pattern in the center of the elytra; this pattern somewhat resembles a sparse fir tree or inkblot test. The pattern typically has very clean borders. Overall elytra shape is slender and narrowed posteriorly. The legs are very long, with white setae densely covering the femora. The lower legs often have red or green metallic highlights. The abdomen is dark brown.

Similar Species Most similar to the Moustached Tiger Beetle (*E. hirtilabris*), but the two can be differentiated by range and physical characters apparent through close-focusing binoculars or camera. The Moustached Tiger Beetle differs by having a dark elytral pattern with irregular borders and by having a labrum with more than 20 decumbent setae. Other mostly white tiger beetles are separated by range or habitat. Ghost Tiger Beetles (*E. lepida*) occur to the north and west of the Whitish Tiger Beetle range and have a rounder elytral shape and pale legs; Eastern Beach Tiger Beetles (*Habroscelimorpha dorsalis* sspp.) are restricted to coastal beaches and bay shores. On both the White Sand Tiger Beetle (*E. wapleri*) and White-cloaked Tiger Beetle (*Eunota togata*), the dark center of the elytra is larger in area than the whitish markings; the opposite is true in Whitish Tiger Beetle.

Habitat Very dry sandy coastal plains habitats, usually in white or very light-colored sugar sand. This sand habitat often has dark flecks in it, which helps this beetle stay camouflaged.

Note This mostly white species can be very difficult to locate on white sand and makes only very short escape flights when disturbed. The areas of coastal plains white sand habitat between the Carolinas and Alabama from which the Whitish Tiger Beetle is absent are mostly occupied by the Moustached Tiger Beetle.

Coastal Tiger Beetle

Ellipsoptera hamata (Audouin and Brullé)

Size 9–13 mm.

Southeast Status Fairly common along immediate Gulf Coast from Mississippi to south peninsular Florida.

Identification A medium-sized brown to very dark brown, occasionally greenish-brown, tiger beetle with extensive elytral markings. The dorsal surface of the head and pronotum are sparsely to fully covered with decumbent white setae. The sides of the head, thorax and abdomen are also densely setose. The labrum is light brown. The right mandible of males lacks a distinct ventral tooth (observation requires in-hand examination or substantial magnification in the field). The elytra are somewhat granulate, typically with extensive, relatively thin cream-colored markings. When complete, the humeral lunule is strongly curved medially with expanded spots at both ends, resembling a capital G. Rarely this lunule is slightly broken, isolating one or both spots. The middle band, if complete, is long, thin, and expanded medially, as is typical for this genus. From the marginal line, it curves

Below: Flight season of the Coastal Tiger Beetle (*Ellipsoptera hamata* sspp.).

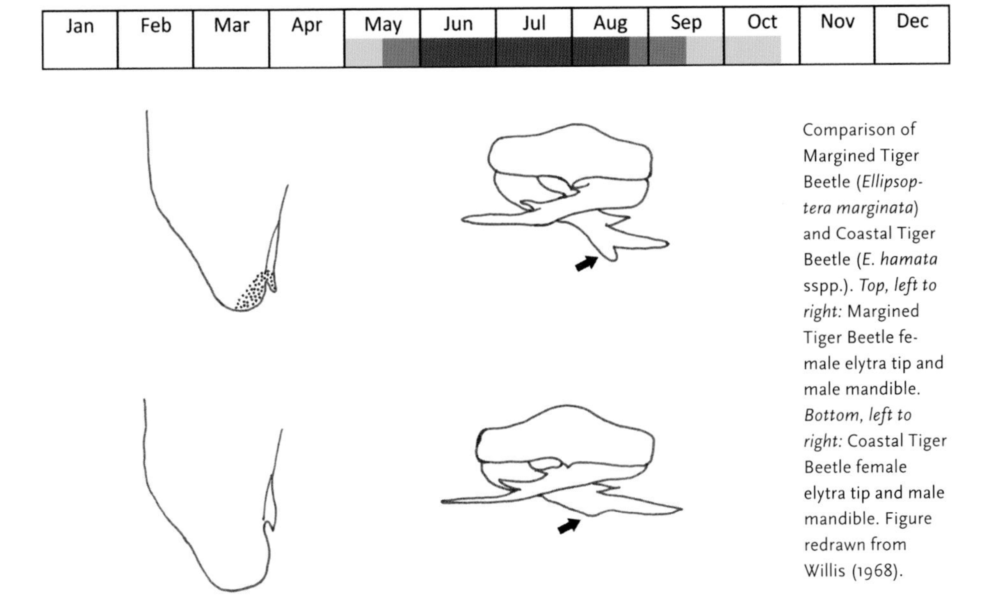

Comparison of Margined Tiger Beetle (*Ellipsoptera marginata*) and Coastal Tiger Beetle (*E. hamata* sspp.). *Top, left to right:* Margined Tiger Beetle female elytra tip and male mandible. *Bottom, left to right:* Coastal Tiger Beetle female elytra tip and male mandible. Figure redrawn from Willis (1968).

forward toward the posterior portion of the humeral lunule before turning sharply rearward and expanding greatly into an area of diffuse wavy lines somewhat resembling a kidney or pea pod in shape. The apical lunule is typically complete and also extends anteriorly along the elytral suture. The marginal line varies in thickness and is almost always complete except for the area between the anterior portion of both humeral lunules. Breaks between major portions of the elytral markings may be present. Occasionally individuals will have all interior markings absent, leaving only the marginal line encircling the elytra. On females, the posterior tips of the elytra are rounded but not deflexed downward, with conspicuous apical spines pointed posteriorly and usually visible in dorsal view. The legs are dark and often have reddish or, less often, greenish metallic highlights. The abdomen is reddish brown.

There are two named subspecies of Coastal Tiger Beetle in the region: *E. h. lacerata* (Chaudoir) and *E. h. monti* (Vaurie). The former subspecies occurs throughout the region and is certainly the form most often encountered here. The latter subspecies is thought to occur in the western part of the species' range and might extend into coastal Mississippi and Alabama. However, the primary feature used to distinguish the subspecies *E. h. monti* is that the elytra are greenish, and this character is highly variable. It is possible that these subspecies are not valid.

Similar Species The Margined Tiger Beetle (*E. marginata*) is extremely similar, but each sex can be separated with careful examination of the following features. In males, the Coastal Tiger Beetle's

Left to right:

Coastal Tiger Beetle (*Ellipsoptera hamata monti*). Mobile County, Alabama.

Coastal Tiger Beetle (*Ellipsoptera hamata lacerata*). Mobile County, Alabama.

Clockwise from top left:

Coastal Tiger Beetle (*Ellipsoptera hamata lacerata*). Baldwin County, Alabama.

Weakly-marked Coastal Tiger Beetle (*Ellipsoptera hamata lacerata*). Dixie County, Florida.

Female Coastal Tiger Beetle (*Ellipsoptera hamata lacerata*). Mobile County, Alabama.

right mandible has only a rounded ventral bump, while the Margined Tiger Beetle has a distinct ventral tooth. In females, the posterior elytral tips are different: the Coastal Tiger Beetle has rounded but not deflexed tips and a greatly retracted apical spine, whereas in the Margined Tiger Beetle the posterior tips of the elytra are rounded and strongly deflexed downward, with inconspicuous apical spines pointed ventrally and not usually visible. From the dorsal view the tips are deflexed downward, and the apical spine is much less retracted. The tips of the female elytra are visible in the field with close-focusing binoculars or through a high-quality camera lens, so females can usually be identified along the portion of the Florida Gulf Coast where the two species co-occur. Males are much more difficult to separate unless captured and examined in the hand. Other members of this genus, such as the Sandbar Tiger Beetle (*E. blanda*), Coppery Tiger Beetle (*E. cuprascens*), and Sandy Stream Tiger Beetle (*E. macra*), are also very similar. However, all these species occur along fairly large freshwater streams and small to large rivers, where Coastal Tiger Beetle is not found.

Habitat Coastal mudflats and salt pannes. Often more easily found at higher tides when water levels push the beetles out of vegetation, and at that time they are also more likely to be found on adjacent sandy beaches.

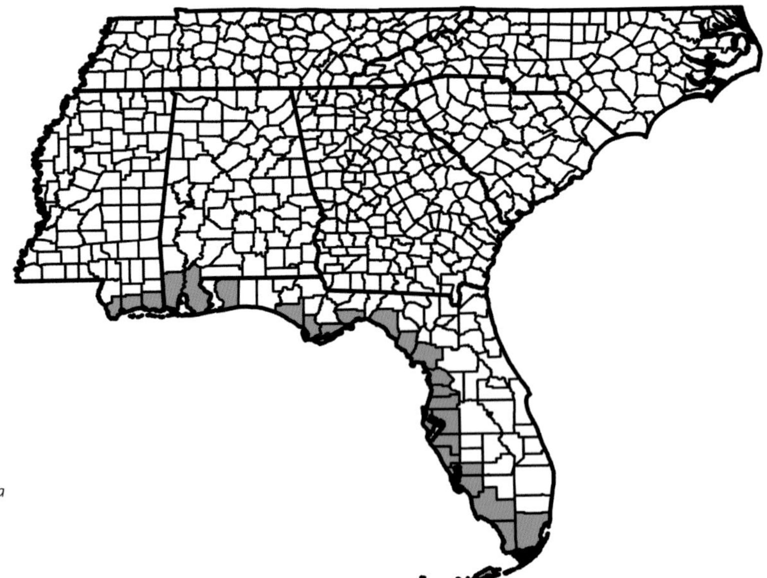

Range of the Coastal Tiger Beetle (*Ellipsoptera hamata* sspp.) in the southeastern United States.

Note This species can sometimes be found on boardwalks adjacent to mudflats when the tide is high. In general, mud flats with lots of fiddler crabs are too wet for this species; like the Margined Tiger Beetle, it prefers the slightly drier adjacent flat areas.

Moustached Tiger Beetle

Ellipsoptera hirtilabris (LeConte)

Size 9–11 mm.

Southeast Status Fairly common in peninsular Florida and southeastern Georgia.

Identification A small to medium-sized pale white tiger beetle. The dorsal surface of the head and pronotum are densely covered with decumbent white setae. The sides of the head, thorax, and abdomen are also densely setose. Setae are sometimes absent or thinner on the dorsolateral edges of the thorax, giving the appearance of two very thin dark lines due to their underlying dark color. The labrum is ivory with many (more than 20) decumbent setae. White maculations are greatly

Below: Flight season of the Moustached Tiger Beetle (*Ellipsoptera hirtilabris*).

Jan	Feb	Mar	Apr	May	Jun	Jul	Aug	Sep	Oct	Nov	Dec

Clockwise from top left:

Pair of Moustached Tiger Beetles (*Ellipsoptera hirtilabris*). Hardee County, Florida.

Moustached Tiger Beetle (*Ellipsoptera hirtilabris*). The ant attached to its antennae is likely the result of a failed predation attempt. Emanuel County, Georgia.

Moustached Tiger Beetle (*Ellipsoptera hirtilabris*). Emanuel County, Georgia.

Close-up of the labrum of a Moustached Tiger Beetle (*Ellipsoptera hirtilabris*). Lowndes County, Georgia.

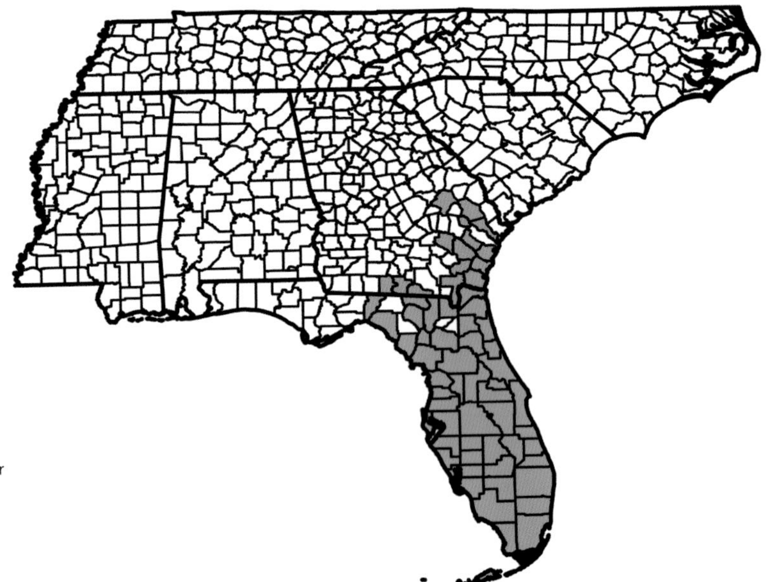

Range of the Moustached Tiger Beetle (*Ellipsoptera hirtilabris*) in the southeastern United States.

expanded, leaving a dark pattern in the center of the elytra. This pattern is variable in shape but is usually widest anteriorly. The pattern typically has irregular borders. The overall elytra shape is slender and narrowed posteriorly. The legs are very long with white setae densely covering the femora. The lower legs often have red or green metallic highlights. The abdomen is dark brown.

Similar Species Most similar to the Whitish Tiger Beetle (*E. gratiosa*), but the two can be differentiated by range and physical characters apparent through close-focusing binoculars or a camera. The Whitish Tiger Beetle differs by having a more cleanly bordered dark elytral pattern and by having a labrum with few or no decumbent setae. Other mostly white tiger beetles are separated by range or habitat. The Ghost Tiger Beetle (*E. lepida*) has a rounder elytral shape and pale legs, and is widely separated by range; Eastern Beach Tiger Beetles (*Habroscelimorpha dorsalis* sspp.) are restricted to coastal beaches and bay shores. On both White Sand Tiger Beetle (*E. wapleri*) and White-cloaked Tiger Beetle (*Eunota togata*), the dark center of the elytra is larger in area than the whitish markings; the opposite is true in the Moustached Tiger Beetle.

Habitat Very dry sandy coastal plains habitat, usually in deep white or very light-colored sugar sand. This sand habitat often has dark flecks in it, which help this beetle stay camouflaged.

Note This mostly white species often does not fly when disturbed and can be very difficult to locate on white sand.

Ghost Tiger Beetle

Ellipsoptera lepida (Dejean)

Size 9–11 mm.

Southeast Status Rare to uncommon and local in Mississippi, western Alabama, and Tennessee. Very rare and local on the upper North Carolina coast and one very old record from central Georgia.

Identification A small to medium-sized tiger beetle. The dorsal surface of the head and pronotum are densely covered with decumbent white setae. The sides of the head, thorax, and abdomen are densely setose. The labrum is ivory. The maculations are greatly expanded so that the elytra appear white, with only small visible areas of underlying brown or green. The overall elytra shape is quite broad for a tiger beetle and is rounded posteriorly. The legs are pale and the femora are covered with white setae. The abdomen is reddish brown.

Similar Species The Ghost Tiger Beetle is the palest member of this genus and has the greatest coverage of expanded maculations; only the Sandbar Tiger Beetle (*E. blanda*) has markings that approach

Below: Flight season of the Ghost Tiger Beetle (*Ellipsoptera lepida*).

Jan	Feb	Mar	Apr	May	Jun	Jul	Aug	Sep	Oct	Nov	Dec

Clockwise from top left:

Ghost Tiger Beetle (*Ellipsoptera lepida*). Stone County, Mississippi.

Ghost Tiger Beetle (*Ellipsoptera lepida*). Perry County, Alabama.

Ghost Tiger Beetle (*Ellipsoptera lepida*). Perry County, Alabama.

Ghost Tiger Beetle (*Ellipsoptera lepida*). Perry County, Alabama.

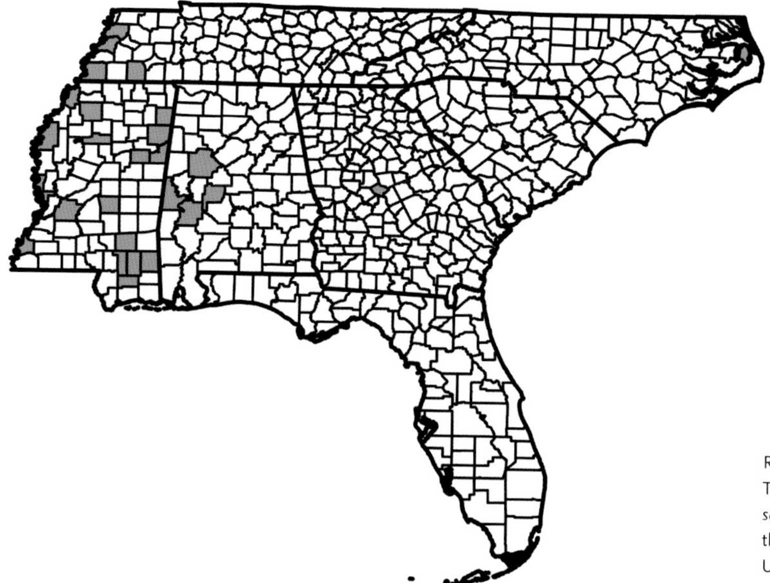

Range of the Ghost Tiger Beetle (*Ellipsoptera lepida*) in the southeastern United States.

being this extensive. The two species are mostly separated by range except in southeastern Mississippi, where the Ghost Tiger Beetle is usually found in very dry habitats away from the water's edge and the Sandbar Tiger Beetle is usually found closer to the water. Additionally, the Sandbar Tiger Beetle has dark legs. The Eastern Beach Tiger Beetle (*Habroscelimorpha dorsalis* sspp.) also has maculations expanded so greatly that the elytra appear almost completely white; however, this species occurs only on coastal and bayside sandy beaches where the Ghost Tiger Beetle is not found (with the exception of the coastal North Carolina record). The Eastern Beach Tiger Beetle's elytra taper to a point posteriorly resulting in a different overall shape, and it has dark legs.

Habitat Very dry sand, usually on the highest portions of large river bars or in sandy areas away from water.

Note The habitat of the Ghost Tiger Beetle gets very hot during the day in the summer, so individuals often burrow into the sand late in the morning for relief. This species makes short escape flights or often simply remains motionless, and its very pale colors render it very difficult to spot on white sand. Recent surveys of the historical site in Georgia have not found this species, and it is believed to be extirpated there.

Sandy Stream Tiger Beetle

Ellipsoptera macra macra (LeConte)

Size 11–14 mm.

Southeast Status Extremely rare; only known from a few specimens along the Mississippi River in west Tennessee.

Identification A medium-sized brown to light brown tiger beetle with distinct elytral markings. The dorsal surface of the head and pronotum are sparsely to entirely covered with decumbent white setae. The sides of the head, thorax, and abdomen are also densely setose. The labrum is ivory to beige. The elytra have pointed apices in males, rounded in females; are weakly granulate; and typically have well-defined white or cream-colored markings. The humeral lunule is usually complete and strongly curved medially with expanded spots at both ends, resembling a capital G. The middle band is long and expanded medially, as is typical for this genus. From the marginal line, it curves forward toward the posterior portion of the humeral lunule

Below: Flight season of the Sandy Stream Tiger Beetle (*Ellipsoptera macra macra*).

Jan	Feb	Mar	Apr	May	Jun	Jul	Aug	Sep	Oct	Nov	Dec

Clockwise from top left:

Female Sandy Stream Tiger Beetle (*Ellipsoptera macra macra*). Eau Claire County, Wisconsin.

Sandy Stream Tiger Beetle (*Ellipsoptera macra macra*). Eau Claire County, Wisconsin.

Sandy Stream Tiger Beetle (*Ellipsoptera macra macra*). Eau Claire County, Wisconsin.

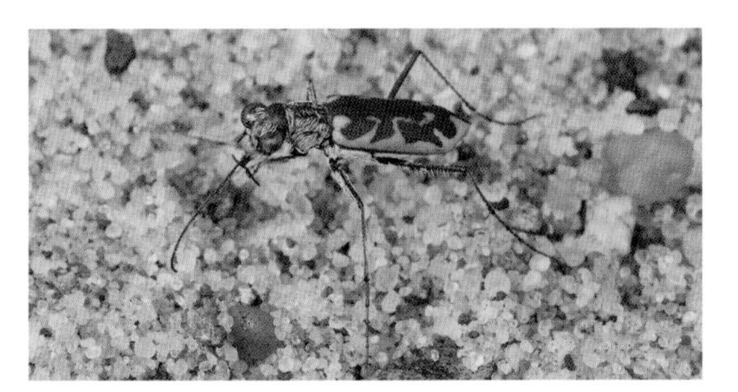

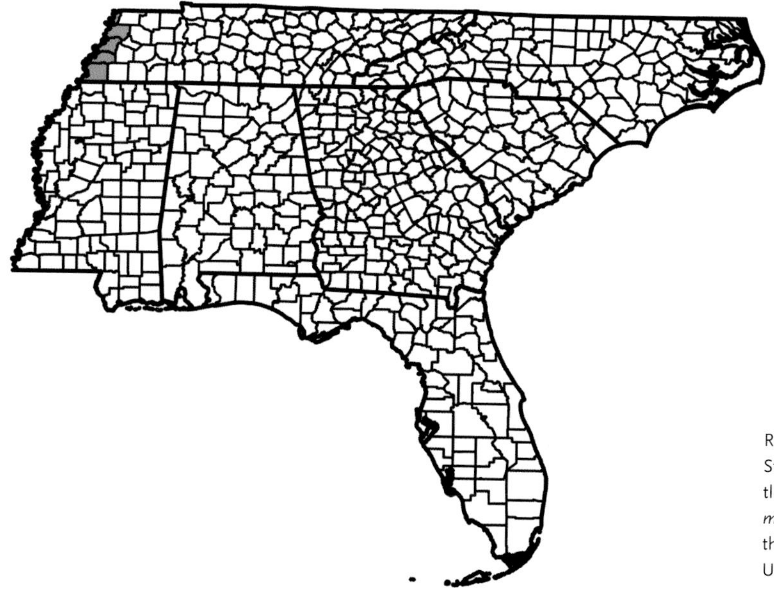

Range of the Sandy Stream Tiger Beetle (*Ellipsoptera macra macra*) in the southeastern United States.

before turning sharply rearward and expanding slightly at the distal tip. Portions of this marking, especially between the rearward turn and distal tip, may be incomplete. The apical lunule is typically complete and also extends anteriorly along the elytral suture. The marginal line varies in thickness and is almost always complete except for the area between the anterior portions of both humeral lunules. The legs are metallic reddish brown, and the femora are usually covered with white setae. The abdomen is reddish brown.

Similar Species The Coppery Tiger Beetle (*E. cuprascens*) is extremely similar, but has more obviously granulate elytra. Additionally, females can be separated by carefully examining the elytra: the apices, which are visible in the field with close-focusing binoculars or through a high-quality camera lens, are rounded in Sandy Stream Tiger Beetles and sharply pointed in Coppery Tiger Beetles. Some sources use the color of the elytra as a distinguishing feature; this character is more helpful when separating more western populations of Coppery Tiger Beetles that are, as the common name suggests, more obviously coppery in color. The more eastern form occurring in the region is less coppery and more brown, like the Sandy Stream Tiger Beetle. Additionally, the Coppery Tiger Beetle is far more likely to be encountered in the region, even along the Mississippi River. Other members of this genus, such as the Coastal Tiger Beetle (*E. hamata* sspp.), Margined Tiger Beetle (*E. marginata*), and Sandbar Tiger Beetle (*E. blanda*), are also very similar. However, the Coastal and Margined Tiger Beetles are restricted to coastal shores, where the Sandy Stream Tiger Beetle is not found; the Sandbar Tiger Beetle is mostly found much farther east or south, with the ranges of the two species completely disjunct.

Habitat Streams and rivers of all sizes; sometimes at large lakes.

Note This species is so similar to Coppery Tiger Beetle that even museum records from the Southeast are often mislabeled; literature records may not always be reliable. Some authorities (MacRae 2009) suggest that this species is much more readily attracted to lights at night than Coppery Tiger Beetle.

Margined Tiger Beetle

Ellipsoptera marginata (Fabricius)

Size 9–13 mm.

Southeast Status Fairly common to common along the immediate coast of most of the region, including the entire Atlantic coast and around the tip of peninsular Florida to the eastern edge of the panhandle.

Identification A medium-sized brown to very dark brown, occasionally greenish-brown, tiger beetle with extensive elytral markings. The dorsal surface of the head and pronotum are sparsely to fully covered with decumbent white setae. The sides of the head, thorax, and abdomen are also densely setose. The labrum is light brown. The right mandible of males has a distinct ventral tooth (see Coastal Tiger

Below: Flight season of the Margined Tiger Beetle (*Ellipsoptera marginata*).

Jan	Feb	Mar	Apr	May	Jun	Jul	Aug	Sep	Oct	Nov	Dec

Clockwise from top left:

Margined Tiger Beetle (*Ellipsoptera marginata*). Glynn County, Georgia.

Close-up of Margined Tiger Beetle (*Ellipsoptera marginata*) female elytra tips. Dixie County, Florida.

Margined Tiger Beetle (*Ellipsoptera marginata*). Charleston County, South Carolina.

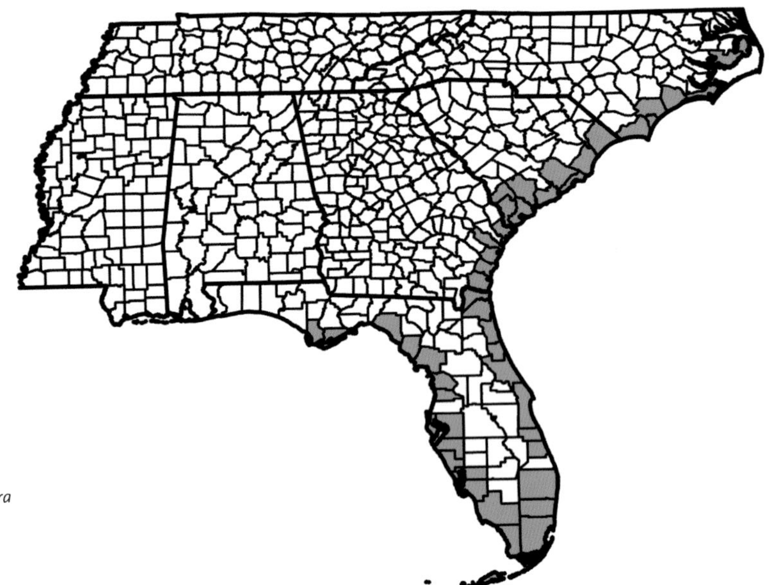

Range of the Margined Tiger Beetle (*Ellipsoptera marginata*) in the southeastern United States.

Beetle diagram; requires in-hand examination or substantial magnification in the field). The elytra are somewhat granulate, typically with extensive, relatively thin cream-colored markings. When complete, the humeral lunule is strongly curved medially with expanded spots at both ends, resembling a capital G. Rarely this lunule is slightly broken, isolating one or both spots. The middle band, if complete, is long, thin, and expanded medially, as is typical for this genus. From the marginal line, it curves forward toward the posterior portion of the humeral lunule before turning sharply rearward and expanding greatly into an area of diffuse wavy lines somewhat resembling a kidney or pea pod in shape. The marginal line varies in thickness and is almost always complete except for the area between the anterior portions of both humeral lunules. The apical lunule is typically complete and also extends anteriorly along the elytral suture. Breaks between major portions of the elytral markings may be present. Rarely individuals will have all or most interior markings absent, leaving only the marginal line encircling the elytra. In females the posterior tips of the elytra are rounded and strongly deflexed downward, with inconspicuous apical spines pointed ventrally and not usually visible in dorsal view (see Coastal Tiger Beetle diagram). The legs are dark and often have reddish or, less often, greenish metallic highlights. The abdomen is reddish brown.

Similar Species The Coastal Tiger Beetle (*E. hamata* sspp.) is extremely similar, but each sex can be separated with careful examination of the following features. In males, the Coastal Tiger Beetle's right mandible has only a rounded ventral bump, while the Margined Tiger Beetle has a distinct ventral tooth. In females, the posterior elytral tips are different: in the Coastal Tiger Beetle the elytra's posterior tips are rounded but not deflexed downward, with conspicuous apical spines pointed posteriorly and usually visible from the dorsal view; in the Margined Tiger Beetle the tips are strongly deflexed downward, with inconspicuous apical spines pointed ventrally (see Coastal Tiger Beetle diagram). The tips of the female elytra are visible in the field with close-focusing binoculars or through a high-quality camera lens, so females can usually be separated along the portion of the Florida Gulf Coast where the two species co-occur. Males are much more difficult to separate unless captured and examined in the hand. Other members of this genus, such as the Sandbar Tiger Beetle (*E. blanda*), Coppery Tiger Beetle (*E. cuprascens*), and Sandy Stream Tiger Beetle (*E. macra*), are also very similar. However, all these species occur along fairly large freshwater streams and small to large rivers, where the Margined Tiger Beetle is not found.

Habitat Coastal mudflats and salt pannes. Often more easily found at higher tides when water levels push the beetles out of vegetation, and at that time they are also more likely to be found on adjacent sandy beaches.

Note This species can also be found on boardwalks or hard packed trails adjacent to mudflats when the tide is high. In general, mud flats with lots of fiddler crabs are too wet for this species; like the Coastal Tiger Beetle, it prefers the slightly drier adjacent flat areas.

White Sand Tiger Beetle

Ellipsoptera wapleri (LeConte)

Size 9–11 mm.

Southeast Status Uncommon in much of Mississippi, extreme south Alabama, and western Florida. Several historical sites in southwestern Georgia, but the species has not been recorded in the state since 1965.

Identification A small, dark brown tiger beetle with limited elytral markings in comparison with most of the beetles in this genus. The dorsal surface of the head and pronotum are sparsely covered with decumbent white setae. The sides of the head, thorax, and abdomen are densely setose. The labrum is ivory. The elytral markings are dominated by a wide marginal line that connects the bases of fairly limited white maculations. The humeral lunule is usually complete, but only the tips extend medially beyond the marginal line, sometimes with

Below: Flight season of the White Sand Tiger Beetle (*Ellipsoptera wapleri*).

Jan	Feb	Mar	Apr	May	Jun	Jul	Aug	Sep	Oct	Nov	Dec

Clockwise from top left:

White Sand Tiger Beetle (*Ellipsoptera wapleri*). Baldwin County, Alabama.

White Sand Tiger Beetle (*Ellipsoptera wapleri*). Stone County, Mississippi.

Pair of White Sand Tiger Beetles (*Ellipsoptera wapleri*). Stone County, Mississippi.

White Sand Tiger Beetle (*Ellipsoptera wapleri*). Okaloosa County, Florida.

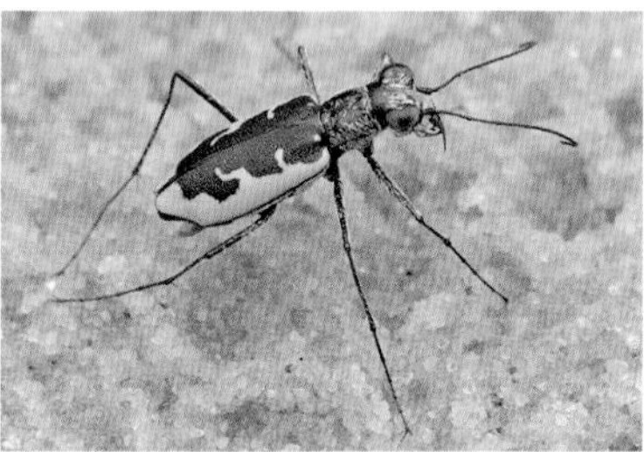

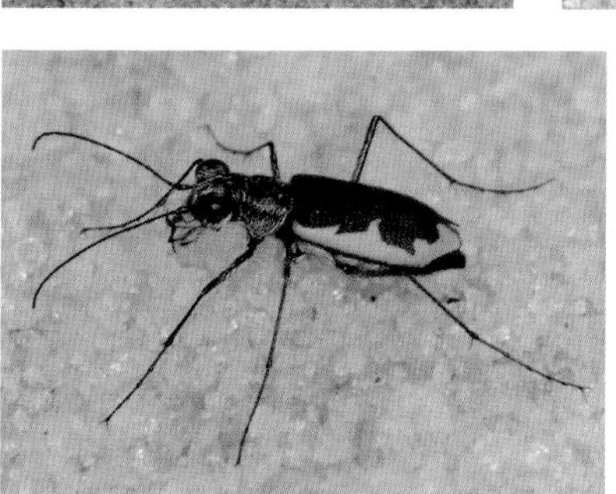

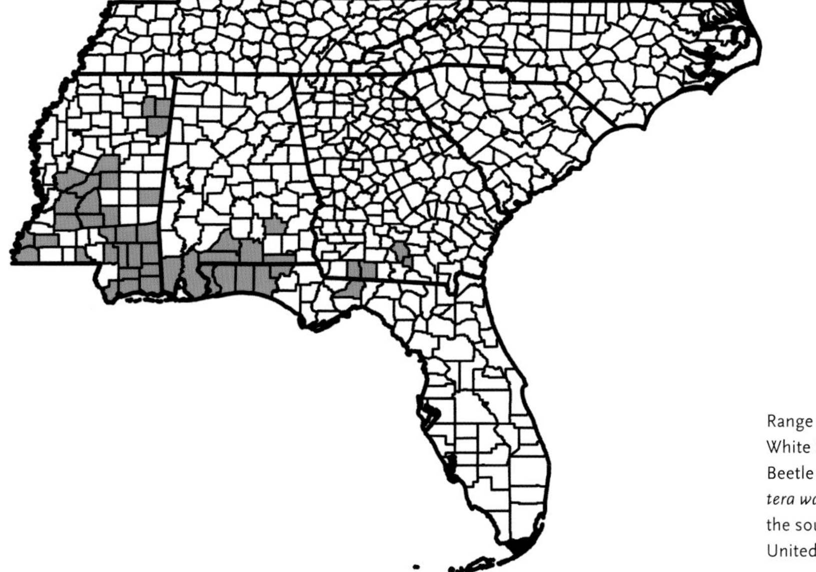

Range of the White Sand Tiger Beetle (*Ellipsoptera wapleri*) in the southeastern United States.

expanded spots at both ends. Middle band is typically complete but thin and short, often with a widened base. The apical lunule is complete but almost completely obscured by the thickened marginal line. The marginal line is complete except for the area between the anterior portions of both humeral lunules. Some individuals have medial elytral markings so reduced that only the marginal line remains. Legs are metallic greenish or reddish brown, and the femora are sometimes covered with white setae. Abdomen is greenish-brown.

Similar Species The Sandbar Tiger Beetle (*Ellipsoptera blanda*) often co-occurs with the White Sand Tiger Beetle in the south of its range, but the White Sand Tiger Beetle is smaller and darker with much less developed elytral markings. The White-cloaked Tiger Beetle (*Eunota togata*) looks similar but has less extensive medial elytral markings and is never found along freshwater streams, only on mud flats and salt pannes along the coast.

Habitat Sandy streams and small rivers, including blackwater streams.

Note Much smaller and less wary than the Sandbar Tiger Beetle, this beetle is usually found along the water's edge and makes only short escape flights.

Gulfshore Tiger Beetle

Eunota pamphila (LeConte)

Size 10–13 mm.

Southeast Status Uncommon and local along the Gulf Coast in Alabama and Mississippi; probably limited in the Southeast due to the lack of appropriate habitats.

Identification A medium-sized pale coppery brown or greenish tiger beetle with strong elytral maculations. The head and pronotum have coppery or greenish reflections. The dorsum of the head is bare; the pronotum has numerous white setae. The sides of the head, thorax, and abdomen also densely setose. Cream-colored elytral markings are expanded and coalesce into wide marginal lines that include medial bulges at the humeral and apical lunules. The middle band

Jan	Feb	Mar	Apr	May	Jun	Jul	Aug	Sep	Oct	Nov	Dec

Above: Flight season of the Gulfshore Tiger Beetle (*Eunota pamphila*).

Top to bottom:

Gulfshore Tiger Beetle (*Eunota pamphila*). Mobile County, Alabama.

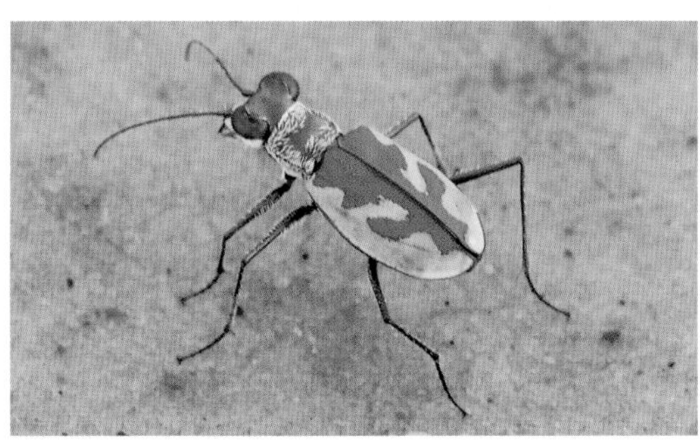

Gulfshore Tiger Beetle (*Eunota pamphila*). Mobile County, Alabama.

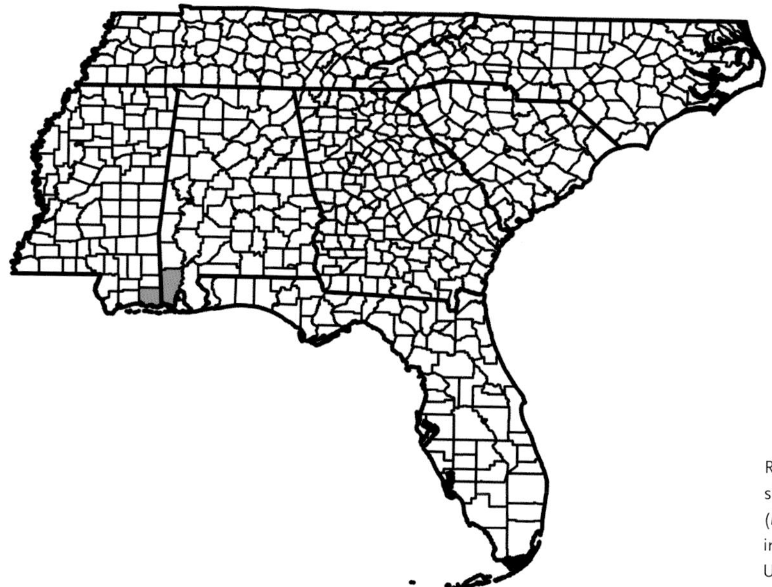

Range of the Gulfshore Tiger Beetle (*Eunota pamphila*) in the southeastern United States.

has a wide, posteriorly angled transverse segment that usually terminates in an expanded spot. The elytra are relatively wide, giving this species a broad appearance. The legs are long and brown with coppery or greenish highlights; the abdomen is greenish brown. Males are noticeably smaller than females.

Similar Species This species may be confused with the White-cloaked Tiger Beetle (*Eunota togata*). The Gulfshore Tiger Beetle is slightly larger and broader in appearance, with a more extensive middle band. The Coastal Tiger Beetle (*Ellipsoptera hamata* sspp.) also occurs in this habitat but is usually darker and has very differently shaped elytral maculations.

Habitat Gulf Coast salt pannes in the region; also found along the Gulf of Mexico shoreline beaches to the west.

Note This somewhat wary species generally prefers to run away rather than fly. It often runs slowly away from the observer but almost always stays out of camera or net reach. Its escape flight is long but slow and easy to follow; it usually turns to face the observer upon landing. It may be numerous in high-quality habitat.

Saltmarsh Tiger Beetle

Eunota severa (LaFerté-Sénectère)

Size 12–15 mm.

Southeast Status Uncommon and local along the Gulf Coast.

Identification A medium-sized dark tiger beetle with few markings. The dorsal surfaces are uniformly green, reddish brown, or flat black; green and reddish-brown forms are more common in Florida, while the black form is more common from Alabama to the west. The labrum is white in males and darker white, ivory, or occasionally brownish in females. The sides of the pronotum are usually covered with white setae, and the abdomen is always densely setose. The moderately granulate elytra are marked with white or cream as follows: The humeral lunule is typically absent in the black form and is either absent or reduced to a single posterior dot in the green or reddish-brown

Jan	Feb	Mar	Apr	May	Jun	Jul	Aug	Sep	Oct	Nov	Dec

Above: Flight season of the Saltmarsh Tiger Beetle (*Eunota severa*).

Top to bottom:

Saltmarsh Tiger Beetle (*Eunota severa*). Mobile County, Alabama.

Saltmarsh Tiger Beetle (*Eunota severa*). Mobile County, Alabama.

forms. The middle band is reduced to a single spot that is occasionally slightly elongated medially, and the marginal line is absent. The apical lunule is thin but complete. The legs are metallic brown or red (rarely green), and the abdomen is black. Individuals from the Florida Keys and southwest Florida are slightly smaller than those from the rest of the species' range.

Similar Species In coastal habitat, the only similar species is the Elusive Tiger Beetle (*E. striga*). That species is dark brown or shiny black rather than flat black, green, or reddish brown. The Elusive Tiger Beetle also has a more cylindrically shaped pronotum, less noticeable elytral markings, usually a brownish labrum, and two obvious rows of punctures paralleling the elytral suture.

Habitat Coastal salt pannes and mud flats. Tolerates a wide range of vegetation density. Can be found near the edges of open flats as well as in small openings along paths and trails through marsh.

Note The Saltmarsh Tiger Beetle is less active than the other species that are found on coastal mud flats and salt pannes, and so it may be more difficult to find. Can be very wary, and often runs to the edge

Clockwise from top left:

Saltmarsh Tiger Beetle (*Eunota severa*). Pinellas County, Florida.

Saltmarsh Tiger Beetle (*Eunota severa*). Dixie County, Florida.

Pair of Saltmarsh Tiger Beetles (*Eunota severa*). Dixie County, Florida.

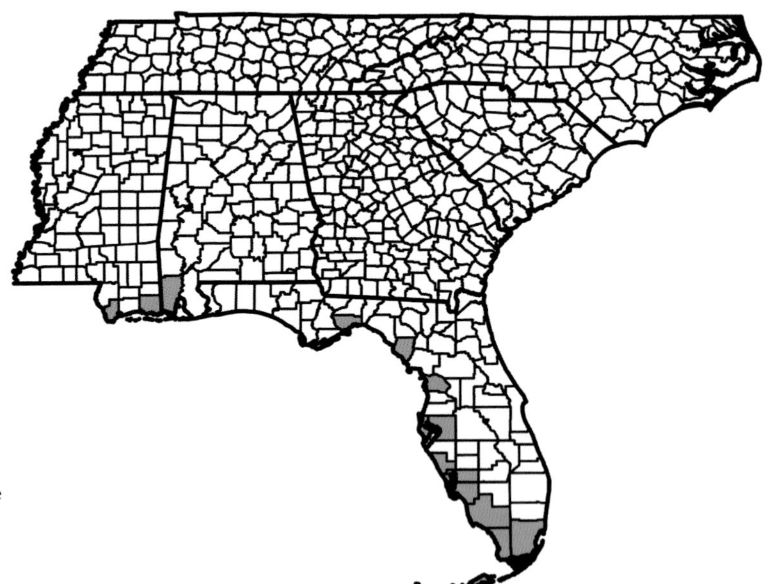

Range of the Saltmarsh Tiger Beetle (*Eunota severa*) in the southeastern United States.

of vegetation rather than making an escape flight, which is typically quite long. This species is most often encountered just after dawn and just before dusk, and it is often attracted to artificial lights at night; as a result, some workers speculate that the Saltmarsh Tiger Beetle may be largely nocturnal.

Elusive Tiger Beetle

Eunota striga (LeConte)

Size 13–17 mm.

Southeast Status Poorly known but apparently uncommon to rare along both Florida coasts and extirpated elsewhere; more records from the Gulf Coast than the Atlantic Coast.

Identification A slender, medium-sized dark brown to black tiger beetle with reduced elytral markings. The head and pronotum have metallic green and coppery reflective areas; green highlights are often extensive. The labrum is typically light in males and darker in females. The pronotum is almost cylindrical; the sides of the thorax and abdomen have few setae. The moderately granulate elytra are shiny and reflective, with a conspicuous row of punctures paralleling the suture on each elytron. The humeral lunule is absent. The middle band is reduced to a single spot, which is rarely extended medially. There is no marginal line. The apical lunule is confined to the elytral margin

Below: Flight season of the Elusive Tiger Beetle (*Eunota striga*).

Clockwise from top left:

Elusive Tiger Beetle (*Eunota striga*). Hillsborough County, Florida.

Elusive Tiger Beetle (*Eunota striga*). Hillsborough County, Florida.

Elusive Tiger Beetle (*Eunota striga*). Hillsborough County, Florida.

Elusive Tiger Beetle (*Eunota striga*). Hillsborough County, Florida.

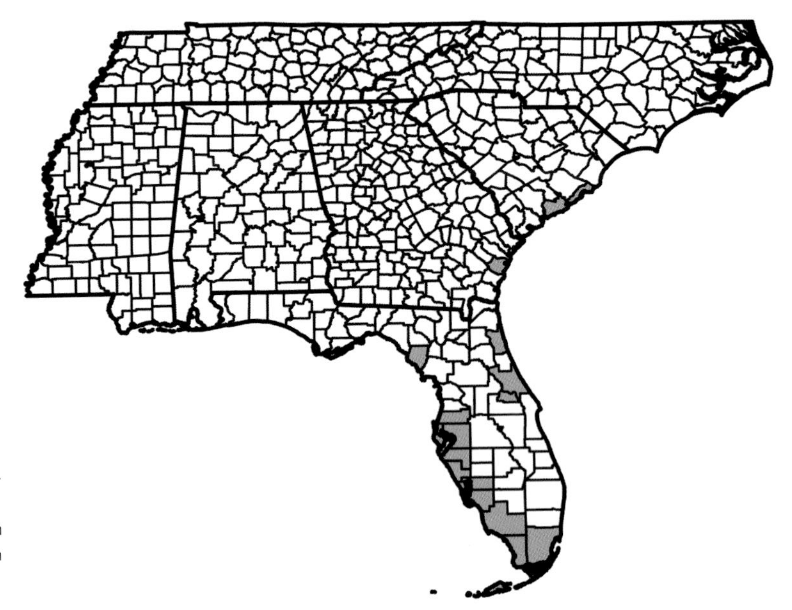

Range of the Elusive Tiger Beetle (*Eunota striga*) in the southeastern United States.

and is often faded, very thin, or both. The legs are dark brown to black with metallic greenish and coppery areas. The abdomen is black.

Similar Species In its coastal habitat, the only similar species is the Saltmarsh Tiger Beetle (*E. severa*). That species is flat black, green, or reddish brown; has more conspicuous elytral markings; does not have a cylindrical pronotum; has a white labrum; and lacks the elytral punctures paralleling the suture.

Habitat Coastal short grass areas with moist substrate; usually near mudflats or brackish canals and waterways.

Note The Elusive Tiger Beetle spends more time within vegetation than any other species in the region and is usually only observed when accidentally flushed into the open or traversing an opening. Forages both during the day and at night but remains within short vegetation. Rarely flies, preferring to run away swiftly and erratically; the occasional escape flights are short and weak.

One Elusive Tiger Beetle specimen was collected in Charleston, South Carolina, in 1924, but there are no other records from the state. A reported record from Georgia has never been verified, and the species apparently no longer occurs outside of Florida. Its affinity for vegetation and swift, erratic running style make it one of the most difficult tiger beetles to observe, capture, or photograph.

White-cloaked Tiger Beetle

Eunota togata togata (LaFerté-Sénectère)

Size 9–11 mm.

Southeast Status Uncommon and local on the Gulf of Mexico coastal mud flats and salt pannes; currently known only from areas near the central Florida Gulf Coast and along the Mississippi and Alabama coasts.

Identification A small to medium-sized tiger beetle with distinctively marked elytra. The head and pronotum have some coppery or greenish reflections and scattered dorsal setae. The sides of the head, thorax, and abdomen are densely covered with white setae. All elytral markings are coalesced into a very wide, cream-colored marginal line with an uneven medial border; the metallic bronze or brown interior portions of the elytra are unmarked. The long legs are brown with coppery or greenish highlights, and the abdomen is brown.

Jan	Feb	Mar	Apr	May	Jun	Jul	Aug	Sep	Oct	Nov	Dec

Above: Flight season of the White-cloaked Tiger Beetle (*Eunota togata togata*).

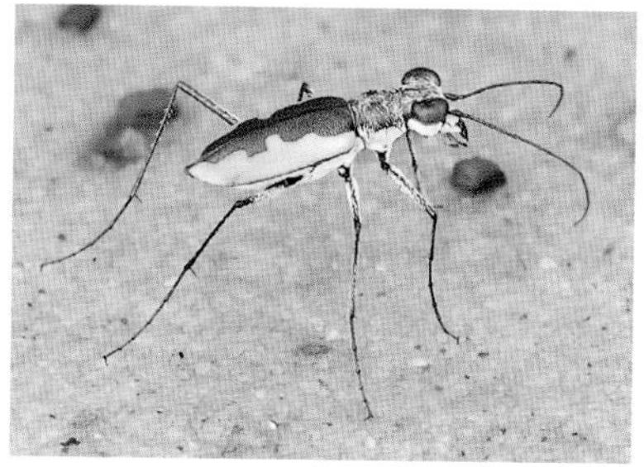

Clockwise from top left:

White-cloaked Tiger Beetle (*Eunota togata togata*). Mobile County, Alabama.

White-cloaked Tiger Beetle (*Eunota togata togata*). Mobile County, Alabama.

White-cloaked Tiger Beetle (*Eunota togata togata*). Dixie County, Florida.

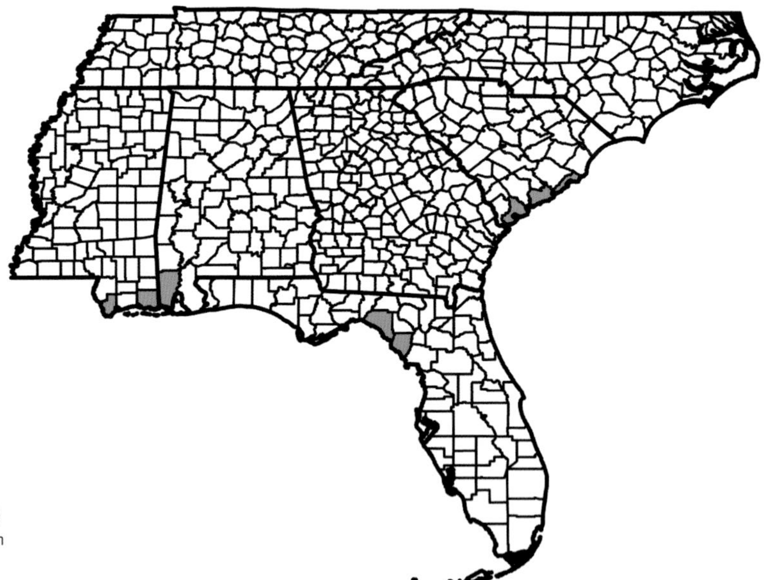

Range of the White-cloaked Tiger Beetle (*Eunota togata togata*) in the southeastern United States.

Similar Species In Alabama and Mississippi this species may be confused with the Gulfshore Tiger Beetle (*Eunota pamphila*). The latter is slightly larger and broader in appearance, with an extensive middle band. The Coastal Tiger Beetle (*Ellipsoptera hamata* sspp.) and Margined Tiger Beetle (*E. marginata*) also occur in this habitat, but are usually darker and always have very differently shaped elytral maculations. The White Sand Tiger Beetle (*E. wapleri*) is similar in appearance but has more extensive medial elytral markings and is restricted to inland freshwater streams.

Habitat Found exclusively on coastal saline mudflats and salt pannes in the region; seeks cover in vegetation but prefers to forage in open areas.

Note This is one of the wariest tiger beetles and is very difficult to approach. Maintaining visual contact with this pale beetle as it makes long escape flights over equally pale sand can be difficult. Usually forages during the warmer part of the day. Numerous specimens were collected from several locations in Charleston and Beaufort Counties, South Carolina, in the early twentieth century; despite extensive searching, the species has not been found there since 1935.

Eastern Beach Tiger Beetle

Habroscelimorpha dorsalis (Say)

Size 8–14 mm.

Southeast Status Uncommon to locally abundant along Gulf of Mexico and Atlantic coastal beaches.

Identification A small to medium-sized pale tiger beetle. This species has maculations expanded so greatly that the elytra are mostly to totally white. There are three subspecies in the region, varying in overall size and degree of expanded maculations.

H. d. media (LeConte) is the largest subspecies in the region (11–14 mm). The head is dark brown with green and coppery reflections, and the labrum is white. The dark brown pronotum may be nearly completely covered with setae, especially laterally; dorsal setae are often less dense or absent. All white maculations on the elytra are greatly expanded, with only a few narrow, mostly longitudinal dark brown markings remaining. The long legs are metallic green or red

Below: Flight season of the Eastern Beach Tiger Beetle (*Habroscelimorpha dorsalis* sspp.).

Jan	Feb	Mar	Apr	May	Jun	Jul	Aug	Sep	Oct	Nov	Dec

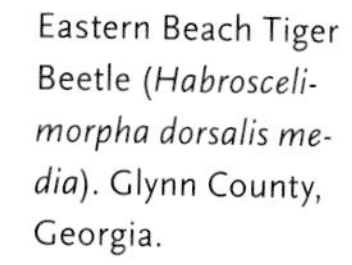

Clockwise from top left:

Eastern Beach Tiger Beetle (*Habroscelimorpha dorsalis media*). Charleston County, South Carolina.

Eastern Beach Tiger Beetle (*Habroscelimorpha dorsalis media*). Charleston County, South Carolina.

Eastern Beach Tiger Beetle (*Habroscelimorpha dorsalis media*). Glynn County, Georgia.

Clockwise from top left:

Eastern Beach Tiger Beetle (*Habroscelimorpha dorsalis saulcyi*). Pinellas County, Florida.

Eastern Beach Tiger Beetle (*Habroscelimorpha dorsalis saulcyi*). Franklin County, Florida.

Eastern Beach Tiger Beetle (*Habroscelimorpha dorsalis venusta*). Mobile County, Alabama.

and frequently have numerous setae along the femora or tibiae. The sides of the reddish-green abdomen are densely covered with setae. This subspecies occurs along the Atlantic Coast from North Carolina to the southern tip of Florida.

H. d. saulcyi (Guérin-Méneville) is intermediate in size for the subspecies in the region (10–12 mm). This subspecies is similar to *H. d. media* with the following exceptions: The dark brown pronotum is usually completely covered with setae, although sometimes less so dorsally. Elytral maculations are even more expanded, so that virtually the entire elytra surface is pearly white with only a very narrow dark brown line along the elytral suture. Rarely, the dark sutural line is slightly expanded with irregular edges. This subspecies occurs along the Gulf Coast from south Florida to Alabama.

H. d. venusta (LaFerté-Sénectère) is the smallest subspecies in the region (8–11 mm) and is essentially a smaller, slightly darker version of *H. d. media*. Elytral markings are virtually identical to that subspecies but with somewhat more pronounced longitudinal dark

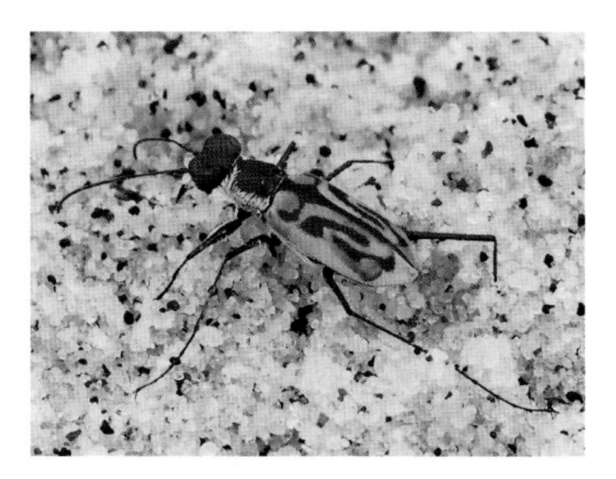

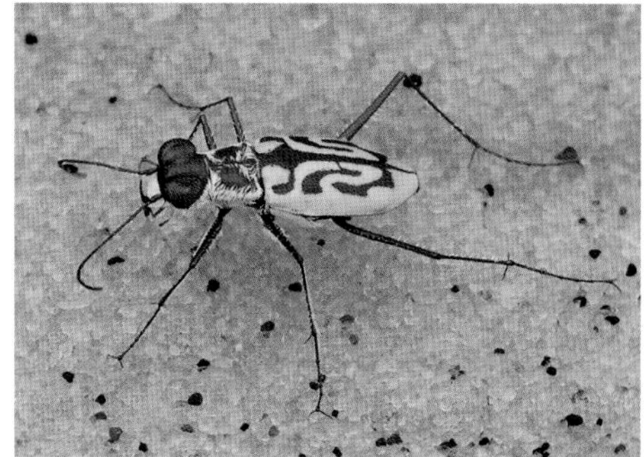

Left to right:

Eastern Beach Tiger Beetle (*Habroscelimorpha dorsalis venusta*). Mobile County, Alabama.

Eastern Beach Tiger Beetle (*Habroscelimorpha dorsalis venusta*). Mobile County, Alabama.

markings. This subspecies occurs along the Gulf Coast from western Florida through Mississippi.

There is much disagreement among tiger beetle experts on the validity of these subspecies, and there may be extensive interbreeding between *H. d. venusta* and *H. d. saulcyi* in western Florida, Alabama, and Mississippi. In all three subspecies, the whitish maculations may become partially cream or brown colored with age.

Similar Species There are no other mostly whitish species on the coastal and bayside sandy beaches where this species is found. There are several whitish species in the genus *Ellipsoptera* in the region, but all are found inland. The lone exception is the Ghost Tiger Beetle (*E. lepida*) along the North Carolina coast in Dare County, where it is found only very rarely on large dunes. The Ghost Tiger Beetle differs from the Eastern Beach Tiger Beetle by having pale legs and broadly rounded posterior elytral margins.

Habitat Coastal sandy beaches along the Atlantic Ocean or Gulf of Mexico. Also occurs on sandy beaches on the mainland side of coastal islands. Usually found along the water's edge or up on drier middle sections of the beach, often foraging near tide debris lines. High tides may force this beetle onto primary dunes or into less sandy areas of beach or mudflat. Heavy recreational use, especially vehicular use, has caused the Eastern Beach Tiger Beetle to disappear from much of its historical range.

Note This species is quite wary and difficult to approach, often running in a zigzag pattern away from the observer before launching into a series of short escape flights. *H. d. venusta*, in particular, often flies

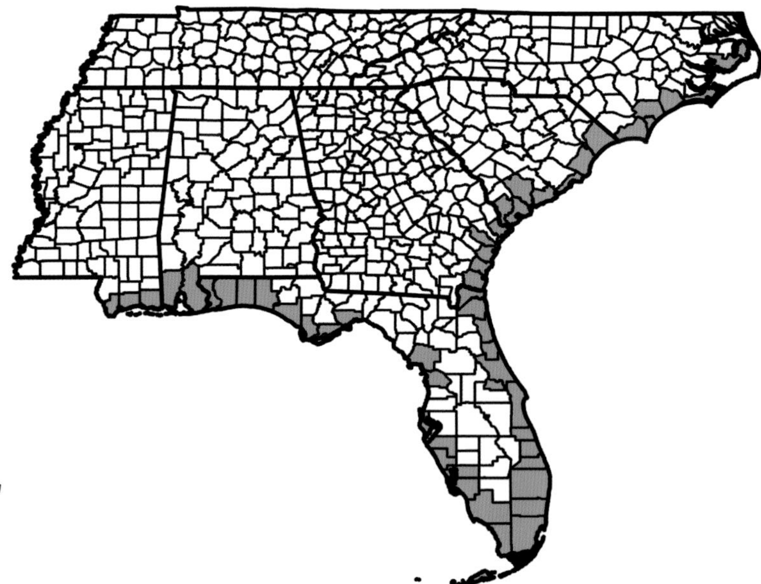

Range of the Eastern Beach Tiger Beetle (*Habroscelimorpha dorsalis* sspp.) in the southeastern United States.

out over the water before looping back to land. A few adults are sometimes active in the cooler morning hours; these individuals are usually easier to approach. Larvae of this species are often active on warm midwinter days; they also employ an unusual method of "wheel locomotion" to escape predators (see Predators and Parasitoids section).

Ant-like Tiger Beetle

Parvindela cursitans (LeConte)

Size 7–8 mm.

Southeast Status Uncommon to rare in western portion of region; populations localized.

Identification A tiny brown or coppery-brown tiger beetle. The labrum is white. The thorax is cylindrical and may be covered with scattered short setae. Elytral maculations are cream colored and fairly complete. The humeral lunule is reduced to a curved posteromedial mark. The middle band may or may not have thin transverse segment; the longitudinal segment is always reduced to a posterior spot. The marginal line extends from middle band rearward and joins the complete apical lunule, which typically has a small medial projection

Below: Flight season of the Ant-like Tiger Beetle (*Parvindela cursitans*).

Jan	Feb	Mar	Apr	May	Jun	Jul	Aug	Sep	Oct	Nov	Dec

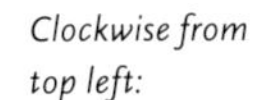

Clockwise from top left:

Ant-like Tiger Beetle (*Parvindela cursitans*). Perry County, Alabama.

Ant-like Tiger Beetle (*Parvindela cursitans*). Perry County, Alabama.

Ant-like Tiger Beetle (*Parvindela cursitans*). The index fingernail in the image provides a size comparison for this tiny species. Perry County, Alabama.

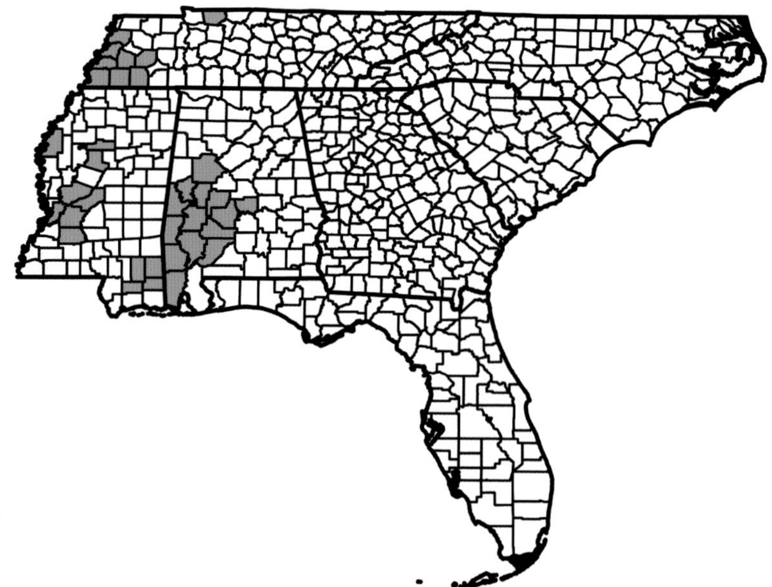

Range of the Ant-like Tiger Beetle (*Parvindela cursitans*) in the southeastern United States.

at its anterior end. The legs are pale coppery brown, and the abdomen is metallic green.

Similar Species Distinctive; no other tiger beetle in the region is this small. More likely to be mistaken for an ant.

Habitat Found on upper portions of river sandbars near vegetation and along sandy trails into riparian forest.

Note Typically stays very close to vegetation, and is often absent during midday. These behaviors and its small size make the Ant-like Tiger Beetle very difficult to locate. Once found, it is not very wary. Not known to fly.

Appendix

Included in this appendix are two species that have been recorded from the southeastern United States but are not currently extant in the region.

Cuban Green-necked Tiger Beetle

Brasiella viridicollis (Dejean)

Size 8 mm.

Southeast Status One individual found on June 4, 1983, in a light trap on Sugarloaf Key, Florida. This is the only record outside of Cuba.

Identification A tiny tiger beetle with a green head and thorax and reddish-brown elytra. This species is unique in the Southeast. Whitish elytral markings typically consist of a humeral dot, a thin middle band connected to a short marginal line, and a complete apical lunule.

Similar Species None in the region.

Habitat In Cuba, bare spots within moist or dry grassy areas.

Season May to September.

Note A weak flier, and conspicuous in flight due to its bright coloration; prefers to run. The means by which this species arrived here is unknown, but it is theorized to be weather related (Schiefer 2004).

Cuban Green-necked Tiger Beetle (*Brasiella viridicollis*), a specimen in Skip Choate's collection, date illegible, collected in Bayamo, Cuba.

Olive Tiger Beetle

Microthylax olivaceus (Chaudoir)

Size 10–12 mm.

Southeast Status Present on the Florida Keys below the Seven-Mile Bridge from about the mid-1940s through the 1980s, but no records since.

Identification A medium-sized brownish-green tiger beetle with elytral markings resembling those of the Bronzed Tiger Beetle (*Cicindela repanda*), but elytra have a strong greenish cast that is absent in the Bronzed Tiger Beetle. The Olive Tiger Beetle is more slender overall with a proportionally longer thorax than the Bronzed Tiger Beetle, which does not occur anywhere near the Florida Keys.

Similar Species None on its coral beach habitat in south Florida.

Habitat Limestone rocky or coral shorelines on the Gulf of Mexico side of islands in the lower Florida Keys; sometimes adjacent to sandy beaches.

Season May to August.

Note It is thought that this Cuban species may have originally been blown up to the Florida Keys via a weather event such as a hurricane; it was established at several sites for about 40 years before disappearing. Most of the known sites have been lost to development (Woodruff and Graves 1963).

Olive Tiger Beetle (*Microthylax olivaceus*), a specimen collected by Skip Choate on June 18, 1977, on Grassy Key, Monroe County, Florida.

Glossary

ABDOMEN The segment of the insect body posterior to the thorax.

AMERICAN DIMINUTIVE TIGER BEETLES Those tiger beetles in the genus *Parvindela*.

AMERICAN TIGER BEETLES Those tiger beetles in the genus *Cicindelidia*.

ANTENNA(E) The threadlike sensory organ(s) attached to the heads of insects.

ANTERIOR Located nearer the front or head.

ANTHROPOGENIC Caused by or the result of human activity.

APICAL Located at the tip of a structure.

APICAL LUNULE The light-colored markings on the apical portion of the elytra of tiger beetles.

APOSEMATIC Conspicuous, usually bright, coloration often serving to indicate some type of chemical defense against predators.

APPALACHIAN PLATEAU Also known as the Cumberland Plateau. The ecoregion located along the western edge of the Appalachian Mountains in Alabama, Georgia, and Tennessee.

ASILIDAE A family of predaceous true flies known as robber flies.

BASKING A behavior that exposes the body surface to the sun in order to obtain heat energy (i.e., to warm up).

BIOGEOGRAPHY The study of a species' distribution through geologic time.

BIOINDICATOR Species that can be used to analyze the effects of changing environmental conditions.

BIOMASS THRESHOLD The amount of energy (food) that must be accumulated by an organism before molting can occur.

Blue Ridge The ecoregion that contains the highest mountains and coolest temperatures in the southeastern United States. Elevations can reach 6,500 feet or higher.

bolus A mass of chewed but undigested food.

Bombyliidae A large family of true flies known as bee flies. The larvae are usually parasitic on other insects including tiger beetles.

Carabidae A large, widespread family of beetles known as ground beetles.

Cicindelidae The family name for the group of beetles referred to as tiger beetles.

Coastal Plain The ecoregion encompassing much of the southeastern United States, including all of Mississippi and Florida, and often typified by sandy soils.

cohort A group of animals of the same species that share a defining characteristic, such as age.

Coleoptera The taxonomic order of insects known as beetles.

complete metamorphosis The type of life cycle in which an insect goes through four distinct developmental stages: egg, larva, pupa, and adult.

contact guarding A type of mate guarding in which the male remains in noncopulatory contact with the female to discourage other males from attempting to mate with the same female.

crepuscular Animals that are most active at dusk and/or dawn.

cuticle The outermost layer that makes up the protective exoskeleton of insects.

disruptive coloration A type of camouflage in which a pattern of boldly contrasting markings serves to break up the overall outline of an organism.

distal Situated away from the body.

dorsal The upper or top side.

ecdysis The process by which an adult tiger beetle emerges from its pupal case.

ecology The relationship of an organism with its environment.

ecoregion A relatively large area of land and/or water that contains a characteristic assemblage of geographically distinct natural communities of plants and animals.

Ellipsed-winged Tiger Beetles Those tiger beetles in the genus *Ellipsoptera.*

elytron (elytra) The hardened forewing(s) of certain insects that serve to cover and protect the delicate flight wings and other parts of the body.

endemic An organism whose occurrence is confined to a particular geographic area.

exoskeleton The tough external covering of the body of certain animals, including many insects.

extirpated A local extinction. The absence of a species from an area where it once occurred.

Fall Line A unique geological feature in the southeastern United States comprised of a line of waterfalls and rapids where rivers pass from the uplands onto the lower, flatter Coastal Plain.

femur The long segment of the insect leg located closest to its body.

genitalia The collective term for the male and female reproductive organs.

genus The taxonomic category used in biological classification that is below a family and above a species. The genus is the first part of a scientific name.

glabrous Without setae.

granulate Grainy or rough in appearance.

Habro Tiger Beetles Those tiger beetles in the genus *Habroscelimorpha.*

holometabolous Insects that undergo complete metamorphosis, with distinct egg, larva, pupa, and adult developmental stages.

humeral Body parts or markings that are located in an area analogous to the shoulder.

humeral lunule The light-colored markings on the humeral region of the elytra of tiger beetles; when complete, these markings are often roughly crescent shaped.

hybridization The mating of two distinct species, resulting in offspring that display a mix of the characteristics of each parent.

IMMACULATE Without spots or markings.

INSECT Any animal in the class Insecta, which are characterized by having the body divided into three distinct regions (head, thorax, and abdomen); possessing three pairs of jointed legs; and (usually) two pairs of wings.

INSTAR An insect that is in a period of growth between successive molts.

INTERIOR LOW PLATEAU An ecoregion that comprises much of middle Tennessee and extreme north-central Alabama and is characterized by limestone sediments and numerous caves.

LABRUM The liplike mouthpart of an insect that covers the oral cavity and most of the other mouthparts.

LAKE WALES SAND RIDGE A ridge of white sands extending north to south for about 150 miles in central Florida; these remnants of prehistoric islands support many distinct species of plants and animals.

LARVA(E) The stage of insects from the time of hatching from the egg until their transformation into the pupal stage, usually including several molts.

LEAF LITTER TIGER BEETLES Those tiger beetles in the genus *Apterodela*.

MACULATIONS The light-colored markings on the elytra of tiger beetles, including the humeral and apical lunules and the middle and marginal bands.

MANDIBLE One of the first pair of insect mouthparts, modified in tiger beetles as a long, sickle-shaped structure used in feeding, mating, and defense.

MARGINAL LINE A maculation running along the outside edge of each elytron, which may or may not be continuous along the entire length of the margin.

MESOTHORAX The middle of the three divisions of the thorax, bearing the first pair of wings and the middle pair of legs.

METALLIC TIGER BEETLES Those tiger beetles in the genus *Tetracha*.

METATHORAX The hindmost of the three divisions of the thorax, bearing the second pair of wings and the third pair of legs.

MIDDLE BAND The light-colored markings on the middle portion of the elytra of tiger beetles. This marking can range from just a spot or two to a complete mark extending from the margin of the elytra to the suture.

MOLTING The act or process of the shedding of the exoskeleton in order to allow for further growth.

MORPHOLOGY The overall form and structure of a living organism.

ORAL CAVITY The area of the insect head located behind the labrum and other mouthparts; it aids in directing chewed and partially liquefied food toward the esophagus.

OVIPOSITION The act of laying eggs.

OVIPOSITOR A specialized structure at the posterior end of the abdomen of some female insects that is used to deposit eggs.

PALP(I) Any of several sensory appendages attached to various mouthparts of certain groups of animals, including insects.

PARASITOID A type of parasite that lives on or in a living host for part of its life cycle and often causes the death of the host.

PHORETIC The description of a nonparasitic relationship between two organisms where one species is carried by the other.

PHYLOGENETIC Related to the development or evolution of a particular group of organisms.

PIEDMONT An ecoregion of the southeastern United States, extending from central Alabama east through north Georgia and much of both South and North Carolina; the Fall Line makes up the southeastern border of this ecoregion.

POSTERIOR Located nearer the rear.

PRONOTUM The dorsal portion of the prothorax.

PROTHORAX The front division of the thorax, bearing the first pair of legs.

PUPA(E) The transformation stage from larva to adult in insects with complete metamorphosis.

RIDGE AND VALLEY Also known as Valley and Ridge. An ecoregion located in northeast Alabama, northwest Georgia, and east Tennessee, and composed of alternating ridges and wide valleys.

RIPARIAN Associated with the area along the bank of a stream, river, lake, or other water body; most often used in association with flowing waters.

Saline Tiger Beetles Those tiger beetles in the genus *Eunota*.

salt panne An area of shallow tidal pools in highly saline brackish marsh areas which host a specific community of plants and animals.

scientific name The name given to a taxonomically unique species of plant or animal based on the system of binomial nomenclature; formed by combining the genus and species names.

sclerite A hardened plate making up part of the exoskeleton of some invertebrates, including insects.

sclerotized Hardened.

seta(e) Stiff hair(s), bristle(s), or bristle-like part(s) of many insects and other animals.

species Regarded as the basic category of biological classification; refers to a group of organisms that share common characteristics and are capable of reproducing fertile offspring.

spermatophore A capsule filled with sperm cells; produced by males of some animals, including insects, and directly transferred to the female's reproductive parts.

spring/fall species Also known as fall/spring species, in reference to tiger beetles that transform into adults during the fall, overwinter underground, and reemerge in the spring to mate and complete their life cycle. They may be quite long-lived as adults.

stilting A thermoregulatory behavior that involves raising the body as high off the ground as possible in order to avoid the higher temperatures of the surface layer of air next to the ground.

subgenus A subdivision of a genus made up of one or more species that share a unique characteristic (or characteristics) that sets them apart from other members of the same genus.

subspecies A taxonomic category that is a subdivision of a species, usually occurring as a result of geographical isolation within a species.

summer species Tiger beetles that emerge as adults from larval burrows in early summer and complete their life cycle in a relatively brief period of time.

sutural line The area along the midline of an adult tiger beetle where the left and right elytra meet.

SUTURE The groove where two sclerites meet.

SYNONYM One of two or more scientific names that have been applied over time to the same species or other taxonomic group.

TARSUS The "foot" of an insect, usually comprising two to five segments.

TAXA Two or more groups within the nested hierarchy of taxonomy. Often used when speaking of species and associated subspecies collectively.

TAXONOMY The science dealing with the identification, naming, description, and classification of living organisms.

TEMPERATE TIGER BEETLES Those tiger beetles in the genus *Cicindela.*

THORAX In insects, the portion of the body that lies between the head and the abdomen.

TIBIA(E) The long segment of the insect leg located between the femur and the tarsus.

TIGER BEETLE A member of a large group of often colorful beetles known for their swift running speed and predatory nature.

TIPHIIDAE A family of insects known as Tiphiid Wasps, whose larvae are parasitoids of certain beetle larvae, including tiger beetles.

TRIDACTYLID A member of the family of insects known as Pygmy Mole Crickets.

VENTRAL The lower or underside.

References

Beaton, G. 2008. Notes on tiger beetle distributions in the state of Georgia, U.S.A., with new county records (Coleoptera: Cicindelidae). Cicindela 40(3):37–45.

Bousquet, Y. 2013. Catalogue of Geadephaga (Coleoptera, Adephaga) of America, north of Mexico. ZooKeys 245:1–1722.

Brzoska, D. W., C. B. Knisley, and J. Slotten. 2011. Rediscovery of *Cicindela scabrosa floridana* Cartwright (Coleoptera: Cicindelidae) and its elevation to species level. Insecta Mundi 0162:1–7.

Choate, P. M., Jr. 1984. A new species of *Cicindela* Linnaeus (Coleoptera: Cicindelidae) from Florida, and elevation of *C. abdominalis scabrosa* Schaupp to species level. Entomological News 95(3):73–82.

Choate, P. M., Jr. 2003. A field guide and identification manual for Florida and eastern U.S.: tiger beetles. University Press of Florida, Gainesville, FL. 197 pp.

Ciegler, J. C. 1997. Tiger beetles of South Carolina (Coleoptera: Cicindelidae). The Coleopterists Bulletin 51(2):177–192.

Duran, D. P., and H. M. Gough. 2019. Unifying systematics and taxonomy: nomenclatural changes to Nearctic tiger beetles (Coleoptera: Carabidae: Cicindelinae) based on phylogenetics, morphology and life history. Insecta Mundi 0727:1–12.

Duran, D. P., and H. M. Gough. 2020. Validation of tiger beetles as distinct family (Coleoptera: Cicindelidae), review and reclassification of tribal relationships. Systematic Entomology (in press).

Erwin, T. L., and D. L. Pearson. 2008. A treatise on the Western Hemisphere Caraboidea (Coleoptera): their classification, distributions, and way of life. Volume II (Carabidae—Nebriiformes 2—Cicindelitae). Pensoft, Sofia, Bulgaria. 365 pp.

Gough, H. M., D. P. Duran, A. Y. Kawahara, and E. F. Toussaint. 2019. A comprehensive molecular phylogeny of tiger beetles (Coleoptera, Carabidae, Cicindelinae). Systematic Entomology 44:305–321.

Grammer, G. L. 2009. A breeding population record of *Cicindela pamphila* in Mississippi and observations on the scavenging behavior of *C. severa* and *C. hamata*. Cicindela 41(3):75–80.

Graves, R. C. 1981. Offshore flight in *Cicindela trifasciata*. Cicindela 13 (3–4):45–47.

Graves, R. C., and D. L. Pearson. 1973. The tiger beetles of Arkansas, Louisiana, and Mississippi (Coleoptera: Cicindelidae). Transactions of the American Entomological Society 99(2):157–203.

Graves, R. C., M. E. Krejci, and A. C. F. Graves. 1988. Geographic variation in the North American tiger beetle, *Cicindela hirticollis* Say, with a description of five new subspecies (Coleoptera: Cicindelidae). The Canadian Entomologist 120(7):647–678.

Guido, A. S., and H. G. Fowler. 1988. *Megacephala fulgida* (Coleoptera: Cicindelidae): a phonotactically orienting predator of *Scapteriscus* mole crickets (Orthoptera: Gryllotalpidae). Cicindela 20(3–4):51–52.

Harvey, A., and S. Zukoff. 2011. Wind-powered wheel locomotion, initiated by leaping somersaults, in larvae of the southeastern beach tiger beetle (*Cicindela dorsalis media*). PLoS ONE 6(3):e17746.

Holt, B. D., and T. W. Barger. 2013. The occurrence of *Habroscelimorpha pamphila* (LeConte) in Alabama. Cicindela 45(1):9–11.

Holt, B. D., and R. S. Krotzer. 2017. Current status of *Cylindera cursitans* (LeConte) in Alabama. Cicindela 49(1):5–9.

Hori, M. 1982. The biology and population dynamics of the tiger beetle *Cicindela japonica* (Thunberg). Physiology and Ecology Japan 19:77–212.

Knisley, C. B. 2011. Anthropogenic disturbances and rare tiger beetle habitats: benefits, risks, and implications for conservation. Terrestrial Arthropod Reviews 4(1):41–61.

Knisley, C. B. 2013. The Highlands Tiger Beetle, *Cicindelidia highlandensis* (Choate): distribution, abundance, biology, and conservation. Cicindela 45(2–3):17–47.

Knisley, C. B., and T. D. Schultz. 1997. The biology of tiger beetles and a guide to the species of the south Atlantic States, Special Publications No. 5. Virginia Museum of Natural History, Martinsville, VA. 209 pp.

Krotzer, R. S. 2013. New records of *Cicindelidia ocellata rectilatera* (Chaudoir) and *Cicindela formosa generosa* Dejean in the Southeastern United States. Cicindela 45(1):1–7.

MacRae, T. C. 2009. The last tiger beetle. Beetles in the bush. Available from: https://beetlesinthebush.com/2009/08/11/the-last-tiger-beetle/ (Accessed 16 July 2020).

Naviaux, R. 2007. *Tetracha* (Coleoptera, Cicindelidae, Megacephalina): Revision du genre et descriptions de nouveaux taxons. Memoires de la Société entomologique de France 7:1–197.

Pearson, D. L., C. B. Knisley, D. P. Duran, and C. J. Kazilek. 2015. A field guide to the tiger beetles of the United States and Canada: identification, natural history, and distribution of the Cicindelidae, 2nd ed. Oxford University Press, New York, NY. 251 pp.

Pearson, D. L., C. B. Knisley, and C. J. Kazilek. 2006. A field guide to the tiger beetles of the United States and Canada: identification, natural history, and distribution of the Cicindelidae. Oxford University Press, New York, NY. 227 pp.

Pearson, D. L., and A. P. Vogler. 2001. Tiger beetles: the evolution, ecology, and diversity of the cicindelids. Cornell University Press, Ithaca, NY. 333 pp.

Schiefer, T. L. 2004. A new record of an endemic Cuban tiger beetle, *Cicindela (Brasiella) viridicollis* (Coleoptera: Carabidae: Cicindelinae), from the Florida Keys. The Florida Entomologist 87:551–553.

Shelford, V. E. 1908. Life histories and larval habits of the tiger beetles (Cicindelidae). Journal of Linnean Society of London, Zoology 30:157–84.

Snodgrass, R. E. 1935. Principles of Insect Morphology. McGraw-Hill, New York, NY. 667 pp.

Stevenson, D. J., G. Beaton, and M. J. Elliott. 2013. The phenology, distribution, habitat, and status of the tiger beetles *Cicindela nigrior* Schaupp and *Cicindela scutellaris unicolor* Dejean (Coleoptera: Cicindelidae) in the coastal plain of Georgia. Cicindela 45(2–3):49–68.

U.S. Fish and Wildlife Service. 2016. Environmental Conservation Online System: Listed Species Summary. Available from: http://ecos.fws.gov/tess_public/reports/box-score-report (Accessed 1 November 2016).

U.S. Fish and Wildlife Service. 2016. Federal Register Vol. 81, No. 194. Available from: https://www.gpo.gov/fdsys/pkg/FR-2016-10-06/pdf/2016-24142.pdf (Accessed 1 November 2016).

U.S. Fish and Wildlife Service. 2016. U.S. Fish and Wildlife Service to List Miami Tiger Beetle as Endangered. Available from: https://www.fws.gov/news/ShowNews.cfm?ref=u.s.-fish-and-wildlife-service-to-list-miami-tiger-beetle-as-endangered-&_ID=35827 (Accessed 1 November 2016).

Vick, K. W., and S. J. Roman. 1985. Elevation of *Cicindela nigrior* to species rank. Insecta Mundi 1(1):27–28.

Willis, H. L. 1968. Artificial key to the species of *Cicindela* of North America north of Mexico (Coleoptera: Cicindelidae). Journal of the Kansas Entomological Society 41:303–317.

Woodruff, R. E., and R. C. Graves. 1963. *Cicindela olivacea* Chaudoir, an endemic Cuban tiger beetle, established in the Florida Keys (Coleoptera: Cicindelidae). The Coleopterists Bulletin 17:79–83.

Xerces Society for Invertebrate Conservation. 2016. Beetles. Available from: http://www.xerces.org/beetles/ (Accessed 1 November 2016).

Illustration Credits

Photographs were provided courtesy of the following:

John C. & Kendra K. Abbott/Abbott Nature Photography, page 3 (left)

Giff Beaton, pages 4, 6 (right), 7 (top), 8, 10, 12 (right), 16, 17 (right), 23–26, 28–32, 36, 38–39, 53 (right), 55, 57 (bottom), 59, 62, 65 (bottom), 67 (bottom), 70, 73 (top right, bottom left and right), 75 (right), 77 (bottom left), 79 (top right, bottom left and right), 82 (bottom right), 83, 84 (top left), 85 (top), 88, 90, 92 (top right, bottom), 94 (bottom), 95, 97, 99, 100 (top), 102 (bottom right), 104 (bottom right), 106 (right), 110 (right), 112 (top left and right, bottom right), 115 (bottom), 118 (bottom), 124 (right, bottom left), 129 (right), 131 (top right, bottom left and right), 134 (bottom), 136, 139 (top, bottom left), 142 (bottom left), 144 (bottom), 147, 149, 151 (left, bottom right), 153, 154 (top), 155, 157 (left), and 161

Wayne Barger, page 13 (top)

Daniel Dye, page 13 (bottom)

Mary Jane Krotzer, 61 (bottom)

R. Stephen Krotzer, pages 2, 3 (right), 6 (left), 12 (left), 17 (left), 27, 35, 53 (left), 57 (top), 61 (top), 64, 65 (top), 67 (top), 69, 73 (top left, bottom middle), 75 (left), 77 (top left and right), 79 (top left), 82 (top right), 84 (top right, bottom), 85 (bottom), 87, 92 (top left), 94 (top), 100 (bottom), 102 (top), 104 (top left and right), 106 (left), 109, 110 (left), 112 (bottom left), 115 (top), 118 (top), 121, 124 (top left), 128, 129 (left), 131 (top left), 134 (top), 139 (bottom right), 142 (top left and right, bottom right), 144 (top), 146, 151 (top right), 154 (bottom), 157 (right), and 160

Ellis Laudermilk, page 102 (bottom left)

Mike Thomas, pages 72, 82 (top left and bottom right)

The diagram on page 1 was drawn by Randy Beaton.

The diagram on page 7 was drawn by Ed Lam.

The diagram on page 127 was redrawn with permission from Willis (1968) by Randy Beaton.

The flight season charts were created by the authors.

The range maps and region map in the introduction were created by Ashley Peters.

Index